S0-ACB-933

High Hopes for High Tech

Urban and Regional Policy and Development Studies

Michael A. Stegman, General Editor

High Hopes for High Tech

Microelectronics Policy in North Carolina

edited by Dale Whittington

The University of North Carolina Press
Chapel Hill and London

Manufactured in the United States of America

Library of Congress Cataloging in Publication Data

Main entry under title:

High hopes for high tech.

(Urban and regional policy and development studies)

Includes index.

1. Microelectronics industry—Government policy—North Carolina—Addresses, essays, lectures.
I. Whittington, Dale. II. Series.

HD9696.A3U5574 1985 338.4'76213817'09756 85-4801

ISBN 0-8078-1667-1

ISBN 0-8078-4138-2 (pbk.)

Contents

Preface

Microelectronics is currently everyone's favorite industry. Considering the stagnant productivity of many traditional sectors of the American economy, it is perhaps only natural that diagnoses of the health of the microelectronics industry command immediate attention from Wall Street analysts, Washington policy-makers, and the media. Touted as the "crude oil of the 1980s," the tiny integrated circuits that the industry produces form the heart of the computer industry, and their uses are expanding throughout the economy. Some analysts believe the world will soon be divided into "have" and "have not" countries according to their ability to participate in this "second industrial revolution"—just the kind of assertion that sets politicians and the public on edge.

The press, Congress, and federal agencies are thus all anxiously watching the competition for shares of the market for integrated circuits for signs that Japan will again succeed in dominating an industry in which United States firms once enjoyed supremacy. The ability of the microelectronics industry to succeed in this competition has become an indicator of American prospects for success in world markets for high-technology products. The Japanese competitive challenge in semiconductors has prompted U.S. firms to lobby for federal support. The Semiconductor Industry Association (SIA), primarily representing the interests of the large "merchant" semiconductor producers that sell integrated circuits to the open market, is seeking public assistance in three areas in order to strengthen the industry's competitive position.

First, SIA has sought to establish competitive parity with Japan in international trade. This pressure has resulted in Japanese agreement to reduce tariffs on semiconductor imports, but success in eliminating nontariff barriers to U.S. exports has proved more elusive. For example, although Japan's publicly owned Nippon Telephone and Telegraph (NTT) did reverse its policy of relying on a favored group of Japanese firms for its sizable procurements of microelectronics products, NTT's large bureaucracy has been slow to solicit bids from U.S. industry.[1] U.S. firms have also been slow in establishing Japanese subsidiaries.

A second objective of SIA is to increase the financial incentives for capital formation to firms in high-technology industries. In fact, that was one of the objectives of the 1981 Economic Recovery Tax Act, one of whose provisions is a 25 percent tax credit for research and development expenditures in excess of those of the prior year.[2] SIA's focus on the industry's capital formation problems reflects the large costs of acquiring sophisticated new machinery to design and manufacture today's highly complex integrated circuits. The need for large capital investments will continue, but profitability is highly cyclical.

The third component of SIA's agenda is a response to the widely shared perception that American university programs are inadequate to meet the needs of today's microelectronics industry, both for technical manpower and for basic research to expand the industry's knowledge base. Microelectronics firms rely heavily on the skills of physicists, chemists, electrical engineers, metallurgists, ceramists, mechanical engineers, industrial and software engineers, computer scientists, and optic specialists, who have been trained at the graduate level. As industry output increases and new products are invented, demand for such specialists generally increases. One estimate is that an additional Ph.D. electrical engineer is required to support each additional $6 million in semiconductor revenues. Because total production by the U.S. microelectronics industry is projected to increase dramatically over the next decade, it is not surprising that analysts and managers are wondering where the manpower to support such growth is going to come from.[3] This concern is heightened by competitive pressure at the international level: Japan is currently producing three times as many electrical engineers per capita as the United States.

Although American universities maintained outstanding programs in microelectronics-related fields during the 1950s and 1960s, many now suffer from unfilled faculty positions and obsolete research equipment. The generous federal support of science education during the post-Sputnik era peaked in 1966 and has been declining in real terms ever since. Because less money is available to support students during graduate training, many potential Ph.D. candidates have preferred to take lucrative jobs in private industry after receiving the B.S. degree. This trend has not only depleted the pool of potential faculty members for future research and teaching, but has also led to constraints on the industry's supply of personnel with advanced degrees. To counter such problems, SIA is pushing for a higher level of public support for scientific training and research.

In addition to this shortage of technical manpower, basic research now being conducted in American universities is insufficient for present and future needs. As the individual transistors and other elements of integrated circuitry become increasingly smaller (the current state of the art is at the scale of one thousandth of a millimeter), it becomes more important for industrial producers to have a better understanding of the electrical and chemical phenomena occurring in very small devices. In some areas technology is overshooting its knowledge base—that is, demands are placed on the technology to do things for which the underlying physical and chemical mechanisms are poorly understood. As a result, industry is becoming more interested in basic research conducted by universities, and in fact SIA recommends closer involvement from both sides, to enhance the flow of information between what are essentially two complementary sources of microelectronics expertise.

The microelectronics industry has become not only a focus of national economic policy but also a target of numerous state and local economic development efforts. Strategies to attract microelectronics firms to specific locales have thus far taken two general forms: (1) the familiar industrial promotion policies that advertise the desirability of a state's "business climate" (often a euphemism for low taxes and lack of union activity), local quality of life, the availability of industrial land and raw materials, publicly funded industrial parks, and low-interest loans, and (2) the economic development strategies specifically directed at the microelectronics industry. States are beginning to assume part of the responsibility for seeking excellence in advanced technologies. California, Arizona, Minnesota, Texas, and North Carolina have all developed microelectronics programs that include substantial increases in public funding for state university research activity in microelectronics and related fields.

Early in 1981 the California legislature approved a $22 million program to protect and invigorate the state's electronics industry, which employs a quarter of the state's manufacturing workers.[4] The largest component of the program is designed to expand basic research in the University of California system; Berkeley, for example, has been awarded $2.6 million to renovate and reequip its microelectronics research facility. Another part of the program, called Microelectronics Innovation and Computer Research Opportunities (MICRO), is designed to encourage joint university–industry cooperation and provide financial support for faculty and graduate students. State funds for MICRO are projected to reach $5 million, all of which is to be matched on an equal basis by private firms (contributions in fiscal year 1982 exceeded $800,000 in cash and $540,000 in new equipment). A third component of the program, the Innovation Research Grant Program, has allocated $400,000 to subsidize applied research at small electronics and other high-technology companies. In addition, $3 million in state funds and $2 million in federal funds are earmarked for grants to inventors and small firms involved in marketing high-technology products. Finally, $5 million in state funds has been reserved for loans to firms seeking to upgrade or expand plant and equipment; this and other services that address capital needs are available for, but not limited to, the microelectronics industry.

Arizona's Center for Excellence in Engineering, at the Arizona State University campus near Phoenix, is a state-funded academic development and training program focused on increasing the number of university graduates trained in fields related to electronics.[5] Though already an important microelectronics center, Arizona had previously lacked a first-class university research program. Local industries have agreed to support the new program with about $10 million in donations. Arizona State University expects to add sixty-eight people to its engineering faculty, and new microelectronics and computer science research facilities are under construction.

In Minnesota, high-technology firms such as Control Data, Sperry, and Honeywell took the initiative in stimulating the interests of state officials in microelectronics.[6] The University of Minnesota already had strong research programs in fields related to microelectronics, and the National Science Foundation had established nearby a national center for surface analysis techniques. University faculty consequently suggested that Control Data support university research in novel techniques in surface analysis that might be applicable to integrated circuit technology. Additional funding from Honeywell and Sperry enabled the creation of the Center for Microelectronics and Information Sciences (MEIS), whose strategy is to leverage private funds with research grants from federal agencies and with matching appropriations from the state legislature. About $7 million in contributions from six private firms is expected to underwrite MEIS operations from 1982 through 1985.

One of the largest state initiatives to date was launched by Texas to attract the Microelectronics and Computer Technology Corporation (MCC), a joint research venture sponsored by twelve major corporate investors, including Control Data, Motorola, Honeywell, National Semiconductor, Digital Equipment, and Advanced Micro Devices.[7] MCC was conceived by William Norris of Control Data Corporation as a private sector response to the microelectronics research program of Japan's Ministry of International Trade and Industry. Texas offered MCC a smorgasbord of inducements—ranging from subsidized mortgages for MCC employees to a job placement service for MCC spouses to free use of a Learjet—that ultimately proved successful; MCC decided in the summer of 1983 to locate in Austin.

A key component of the strategy to attract MCC was a commitment to create world-class programs in computer science and electrical engineering at the University of Texas at Austin and at Texas A&M University. The University of Texas established a $15 million endowment (in addition to an existing one of $5 million) to support faculty positions in these fields and to create thirty new faculty positions in microelectronics and computer science over a period of three years. In addition the University of Texas pledged to provide the following: (1) additional graduate fellowship support in microelectronics and computer science of $750,000 annually; (2) an additional $1 million annually for equipment maintenance, support, technical personnel, and other operating expenses; (3) $5 million, over two years, to acquire capital and laboratory equipment for research; and (4) an office and laboratory building (200,000 square feet) at a nominal rent on land owned by the University of Texas system.

North Carolina has also targeted its industrial development efforts at the microelectronics industry.[8] In 1981 it established an independent, nonprofit corporation, the Microelectronics Center of North Carolina (MCNC), funded by an initial $24.4 million state appropriation. The objective of MCNC, in the words of one of its brochures, is "to develop an educational and research activity in microelectronics that will establish North Carolina as a national center in this significant technology." Duke University in Durham, North Carolina State University in Raleigh, the University of North Carolina at Chapel Hill, and Charlotte, North Carolina A&T University in Greensboro, and the Research Triangle Institute will participate in MCNC activities and, along with four state appointees, be represented on its Board of Directors.

Construction began in 1982 on a new $10.5 million microelectronics research facility, to be located in the Research Triangle, an area of the state that has successfully established itself as an attractive location for high-technology firms. Initial equipment for MCNC will cost $8.6 million, and $2.8 million will be used for operating expenses for the first two years. Of the $2.5 million allocated for program support over two years, part is being used to support graduate students at each of the five universities, and part will go for teaching laboratories to be used by their physics and chemistry departments.

The MCNC facility includes data transmission links that allow the participating universities to share a central technical data bank and to receive MCNC-originated classroom instruction. A contamination-free integrated circuit production facility operating at the highest industrial standards of cleanliness will enable experimental wafer fabrication by university researchers and MCNC staff. Industrial affiliates also have access to the data bank, curriculum transmission, and the wafer fabrication equipment.

Plans for MCNC include close cooperation with the North Carolina community college system. The overall intention is to educate an increased number of scientists, engineers, and technicians in the most advanced microelectronics design and fabrication techniques. This integrated approach to training for the microelectronics workforce, it is hoped, will increase the credibility of North Carolina's efforts to make itself a center for the microelectronics industry. In fact, the establishment of MCNC coincided with General Electric's decision to locate a major microelectronics facility in the Research Triangle.

The papers collected in this volume examine some of the planning and policy issues raised by North Carolina's effort to attract the microelectronics industry to the state. It is still, of course, too early to judge whether the programs in North Carolina or in other states will be successful in terms of their stated objectives. High-technology industrial recruitment is, however, a critical issue in national and state-level economic development today. It is thus useful to examine in some detail the attempt of one state—North Carolina—to establish itself as an upcoming center of the industry. Many of the objectives, underlying assumptions, political ramifications, and social consequences of the state's policies and planning may be of interest not only to North Carolinians, but also to communities throughout the nation.

Although North Carolina's effort is representative of state policies adopted elsewhere, it has two distinguishing features that especially justify closer scrutiny. First, North Carolina is seeking to achieve excellence in microelectronics education within universities where that branch of technology has been relatively neglected. Second, it hopes to establish an industrial base in microelectronics in a region that lacks substantial existing investment in microelectronics. A study of North Carolina's experience thus provides an opportunity to examine the appropriate roles for state policy in such circumstances. It is important to realize, however, that the ingredients of success in international competition do not seem to come primarily from public sector actions in either the United States or Japan. In fact, success periodically eludes the most stellar firms in the industry.[9] Federal, state, and local policies can improve the economic and technological environment within which microelectronics firms operate, but they do not insure success in an industry in which technologies are changing at a dizzying pace.

The public debate over microelectronics policy in North Carolina and many other communities has thus far tended to focus on whether microelectronics companies can be induced to locate facilities in specific areas and what wages they would offer to potential employees. These concerns are perhaps natural given our experience with capital accumulation during previous phases of industrialization. But in the present case it is important to consider not just the manufacture of silicon chips, but the applications of microelectronics technology by other industries as well. Indeed, the adaptation of this technology to various ends (from microcomputers and industrial robots to digital watches and children's games) has itself become an industry. This technological diffusion and diversity in applications makes the formulation of industrial and economic development policy extremely difficult. For example, a recent study that examined the adoption of new technologies such as computers, computerized numerical control systems, and microprocessors in the American machinery industry found that these technologies were being adopted most rapidly in older plants in the industrial Midwest, not in new facilities or those in the Sunbelt.[10] Yet most proponents of a national industrial policy would have us as a nation disinvest in precisely such "sunset" facilities.

A serious examination of high-technology economic development policy must consider not only the potential successes in terms of economic growth and the establishment of institutions, but also the impact on the labor force, environmental consequences, and unforeseen effects of policy. During the time I was editing this volume, I was also investigating economic development problems in the Third World, and had the opportunity to observe some interesting parallels between the high-technology economic development concerns of many state governments in the United States and of national governments of developing countries. Both often focus on attracting "foreign" capital and offer in exchange a variety of financial inducements of often dubious value. Because of their lack of leverage in the international economy, both face a "prisoner's dilemma" situation: each government is forced to adopt economic development strategies to counteract those of its neighbors, a tactic that in the end benefits none of them.

Despite this focus on attracting capital, governments in many developing countries and in many states are equally preoccupied with preventing the departure of existing assets. Both are troubled by the consequences of capital mobility: a company that can bring the benefits of economic growth can also leave—and often does—if corporate interests change. At a 1982 conference on the impacts of the microelectronics industry, the president of Victor Technologies summed up the problem facing communities attempting to exert some control over their future economic development: "I am a world citizen; I sell in world markets." The implication was that communities must adapt to the changing realities of international competition, and not adopt policies that run counter to them.

One of the principal attractions for industry in locating in both the developing countries and the southern United States is low-wage labor, both skilled and unskilled. In addition multinational corporations are often attracted to countries with authoritarian regimes that have the ability to control the local labor force. The appeal is somewhat similar in the southern United States, where state governments have often adopted strong antiunion policies and advertised their low wage scales, nonorganized labor, and good business climate.

Despite these similarities, there is one striking difference between the official economic development philosophy in most developing countries and that of state governments in the United States. In dealing with Third World countries, the United States government, acting directly through the United States Agency for International Development and indirectly through multilateral donors such as the World Bank, acknowledges the need to deliver the benefits of economic development efforts to the "poorest of the poor." Most international donor agencies have an explicit policy emphasis on helping the most underprivileged in society and are willing to give up some gains in economic efficiency to achieve larger objectives of income redistribution. Donor agencies also pay at least lip service to the need to consider the socioeconomic impact of projects—the disruption of communities, the breakdown of traditional ways of life, and the social consequences of technological change. In contrast, high-technology economic development policy in the United States is essentially based on a belief in "trickle-down" economics. It is clear, at least, that the immediate beneficiaries of the rapid expansion of the high-technology sector are the stockholders of the new companies, the individuals with technical skills that are in short supply, and the owners of other resources in areas of rapid growth who are able to capture economic rents. Vast new fortunes have been made; there are an estimated fifteen thousand new millionaires in Silicon Valley alone.

The benefits of economic growth from high technology are, however, assumed to be ultimately good for all in the sense that growth improves the general economy. The products of these new technologies are changing society in a multitude of ways that most of us have only dimly grasped, but they are nonetheless exciting to large segments of society. This "second industrial revolution" will entail major disruptions in our patterns of work and, at least in the short run, may destroy more jobs than it will create. High-technology economic development policy has so far not been particularly concerned with the consequences of this shift in employment patterns, or with the question of precisely whom this economic progress is to benefit. In developing countries, efforts by donors and national governments to tie economic growth policy to equity considerations have yielded mixed results. In debates on high-technology development policy among industry and government participants a fear is often expressed that, whatever the social costs, as a country we simply cannot afford to lose our competitive edge in microelectronics technology. Hence the argument that the public sector must assist industry in achieving excellence in microelectronics technology and deal with the employment and social consequences later.

This volume is divided into three parts. Following the introduction, the first part contains five chapters that present a description of the microelectronics industry and background issues related to policy in high-technology economic development from a national perspective. The second part contains two chapters on questions of high-technology industrial location at the regional level, in North Carolina and in the South Atlantic area. The third part includes five chapters that focus on state-level policy and planning issues related to microelectronics development.

Notes

1. *Business Week*, 9 August 1982, 38–44.

2. Instead of reducing the statutory rate of corporate taxation, the Economic Recovery Tax Act of 1981 (ERTA) shortened depreciation writeoff periods and subsidized certain assets through the investment tax credit. Although the Tax Equity and Fiscal Responsibility Act of 1982 (TEFRA) rescinded nearly 60 percent of the original ERTA tax cut, overall effective corporate tax rates have fallen significantly from 1980 levels. Charles R. Hulten and James W. Robertson, "The Taxation of High Technology Industries," *National Tax Journal* 38 (September 1984): 327–45, argue, however, that these methods of reducing corporate income taxes have not spread tax relief evenly across industries and have effectively penalized high-technology firms relative to more traditional "smokestack" industries. Net benefits to the microelectronics industry in particular as a result of the new depreciation schedules for manufacturing equipment have been small because the previous practice was to write off equipment over its useful life, which in the case of semiconductors averaged four years (*Science*, 21 August 1981; *Wall Street Journal*, 27 February 1981).

It is difficult to predict the impact of the 1984 Treasury tax reform plan (if it were to be enacted) on the growth and spatial distribution of the microelectronics industry. Some of its provisions would further reduce the depreciation and investment tax credit advantages of ERTA and thus raise effective corporate tax rates. To judge from Hulten and Robertson's findings, however, the Treasury tax reform plan might result overall in more equitable treatment of high-technology industries.

3. Mel H. Eklund and William I. Strauss, *Status 1982* (Scottsdale, Ariz.: Integrated Circuit Engineering Corporation, 1982).

4. *New York Times*, 12 February 1981, D1; *Physics Today*, March 1982, 57–58.

5. *Science*, 6 August 1982, 512–13.

6. Nico Hazewindus, *The Microelectronics Capability of the United States* (New York: New York University Center for Science and Technology Policy, 1981), 178–79; Robert Hexter, of the University of Minnesota, remarks at the U.S. Microelectronics Industry and Public Policy seminar, New York University Center for Science and Technology Policy, 2 April 1982.

7. "Cooperation is the Key: An interview with B. R. Inman," *Communications of the Association for Computing Machinery* 26, no. 9 (September 1983).

8. *Physics Today*, March 1982, 57–58.

9. *Fortune*, 9 August 1982, 40–45.

10. John Rees, Ronald Briggs, and Raymond Oakey, "The Adoption of New Technology in the American Machinery Industry," Working Paper, Department of Geography, Syracuse University, 11 July 1983.

High Hopes for High Tech

1 Microelectronics Policy in North Carolina: An Introduction

Dale Whittington

The push to make North Carolina a center of the microelectronics industry began in early 1980. The General Electric Corporation informed Governor James B. Hunt that it was considering North Carolina as a possible location for a new microelectronics research, development, and production plant. General Electric was actually not a technological leader in the microelectronics field. It had dropped out of semiconductor production in the early 1970s, and had just recently reentered the business. Its primary objective was to produce chips for internal use in its products. Nevertheless a $100 million facility is a prize catch for most states these days, and Governor Hunt became heavily involved personally in the recruitment effort. In fact, by all accounts it seems to have been his idea and carried out under his direction. The recruitment of this General Electric Microelectronics Center precipitated a broader set of state policy initiatives designed to make North Carolina a center of the growing microelectronics industry.

Governor Hunt's microelectronics initiative can only be properly understood within the context of past industrial recruitment and economic development policies in North Carolina. During the last fifty years the state has been spectacularly successful in its industrial recruitment efforts. Over the period 1939–1954 the number of manufacturing establishments increased from 3,158 to 7,500—more than any southern state. In 1954 North Carolina ranked first in the nation in the value of textile, furniture, and tobacco manufactures. It also ranked forty-first in per capita income and last in hourly wages in manufacturing. The result of all this industrial recruitment had essentially been the establishment of low-wage branch plants, which still represented an improvement over the prevailing wages in the agricultural sector.

As discussed by Hekman and Greenstein (this volume), the North Carolina economy today is diversifying out of textiles, furniture, and tobacco; for example, the state now has 50,000 jobs in electronics. In the 1970s North Carolina led all Southern states in new manufacturing investments, although these were still largely low-wage, branch plants consistent with the long southern tradition. Today North Carolina still ranks forty-first in per capita income and has the lowest manufacturing wages in the country, 15 percent below the national average. North Carolina also has the highest percentage of its workforce in manufacturing.

The growth of research and development activities in the Research Triangle Park over the past twenty years has stood in stark contrast with the low-wage industrial development occurring elsewhere in North Carolina. The Research Triangle Park has been a symbol of the promise of economic development that has not yet reached most North Carolinians.

The factors that have led to the success of the Research Triangle Park are the starting point of the discussion below, for they are also in a sense the foundation of Governor Hunt's recent initiative of a formal state policy to attract the microelectronics industry to North Carolina. The cornerstone of that initiative, the Microelectronics Center of North Carolina, serves as a focus for the second portion of this chapter, which concerns the evolution and rationale of the new state policy. The likely effects of this policy, specifically on the state's economy but also on the environment and quality of life, have been much debated, and the topic has received a great deal of public attention. The third section of this introduction reviews the major issues in question in light of research reported in this volume, and the concluding remarks offer some observations on what other states may learn from North Carolina's policy and planning for recruiting the microelectronics industry.

The Research Triangle Park of North Carolina

The Research Triangle Park (RTP) is now widely recognized as one of the most successful planned industrial research and development parks in the world. Delegations of government officials and economic development planners from other states and countries now routinely visit RTP in hopes of learning how to replicate its heavy concentration of research and development activities.

There are almost 20,000 employees working in about forty establishments in RTP, and combined annual payroll is now on the order of $.5 billion. Table 1 lists the major facilities in the park, their date of arrival, the current size of their site, and their 1982 employment level. There is a great deal of diversity among the current research operations installed there; RTP cannot be characterized as focused on one or two industrial sectors or problem areas. Five of the establishments are part of the federal government; several are affiliated with the North Carolina state government or universities in the Triangle cities—Raleigh, Durham, and Chapel Hill. The majority of establishments are research and development branches of major private corporations.

Today promoters of RTP and its related activities exude confidence about the future growth of research and development in North Carolina and the Triangle area. This optimism is largely independent of future growth of the microelectronics industry in the region. Indeed, by the mid-1970s RTP had seemingly accumulated a critic mass of research personnel and activities that now makes it much more attractive for prospective occupants than it was fifteen years ago.

This optimism, however, has long been prevalent. In 1965 an article in *Science* stated, "As one looks back, the Triangle has about it an aspect of historical inevitability."[1] It is perhaps natural that the press and the participants in the founding of RTP now look back over the last thirty years of the development of the Research Triangle and sense something inevitable about the economic growth they have witnessed. The foundation for development was certainly in place with the close proximity of three major universities—Duke University in Durham, North Carolina State University in Raleigh, and the University of North Carolina at Chapel Hill—without which every observer agrees that the development of Research Triangle Park could never have happened.[2] Yet in the 1950s none of these institutions was a truly first-class university. All three universities have grown in reputation over the last thirty years, but they have all benefited from the Research Triangle Park perhaps as much as the Research Triangle has benefited from them. The University of North Carolina at Chapel Hill was widely acknowledged as the finest university in the South, but the competition for this honor was not keen. There were certainly twenty other universities in the United States of comparable stature.

Although the Triangle area had three major universities and a pleasant climate, if one were betting in the 1950s on a location for a concentration of research and development activities in the South, the Triangle would have been only one of several candidates. Indeed, Florida and Texas would both have seemed to hold more promise. Florida was the first southern state to experience an influx of more technologically oriented industries, and it benefited significantly from the initial phase of the space program. Florida rapidly accumulated a more technologically advanced industrial base and a larger skilled work force than the rest of the South. (Even today Florida maintains this lead in electronics

Table 1. Occupants of the Research Triangle Park

	Date of Arrival	Size (acres)	1982 Employment
Research Triangle Foundation of North Carolina	1959	170	5
Research Triangle Institute	1959		1,000
Chemstrand Research Center (Monsanto)	1960	104	200
USDA Forest Service*	1962	27	36
American Association of Textile Chemists and Colorists	1964	10	22
North Carolina Science and Technology Research Center	1964	9	30
Triangle Universities Computation Center	1965		39
Technitral, Inc.[a]	1966		
National Center for Health Statistics Laboratory*	1966	15	150
IBM Corporation	1966	408	8,000
National Institute of Environmental Health Sciences*	1966	41	
Beaunit Fibers[b]	1966		
Hercules, Inc.[c]	1968	46	
US Environmental Protection Agency[d]*	1968	49	1,500
Richardson-Merrell, Inc.[e]	1968		
Becton, Dickinson & Co.	1969	53	100
Burroughs Wellcome Co.	1969	100	1,020
Troxler Electronic Lab	1973	27	89
US Army Research Office*	1975	10	97

[a] Moved from RTP in 1973; property sold to Troxler Elextronics

[b] Moved to Raleigh in 1975; building occupied by EPA

[c] Moved and leased building to IBM

[d] Formerly National Center for Air Pollution Control Administration

[e] Purchased land

Table 1, continued

	Date of Arrival	Size (acres)	1982 Employment
International Fertility Research Office	1976	8	99
Chemical Industry Institute of Toxicology	1977	56	103
Triangle Universities Center for Advanced Studies, Inc. (TUCASI)	1977	120	
National Humanities Center*	1977	15[f]	20
Data General Corporation	1977	25	185
Airco	1979		14
J. E. Sirrine Co.	1979	31	210
Northrop	1979		380
Instrument Society of America	1979		61
Mead Environmental Division	1979		100
TRW Environmental Engineering Division	1980	21	80
Northern Telecom, Inc.	1980	73	1,800
General Electric Microelectronics Center	1981	110	160
Union Carbide Corporation	1982	51	450
Glaxo, Inc.	1982	21	
Microelectronics Center of North Carolina[g]	1982	36[f]	12
General Telephone & Electronics	1982[h]		
Sumitomo Electric	1982[h]	34	
Ciba-Geigy Corporation	1982[h]	12	
DuPont Corporation	1983[h]	100	

[f]Part of TUCASI's 120 acres

[g]Under construction

[h]Announced in that year

*Federal government facility

and other high-technology industrial sectors.) Moreover, Florida offered certain advantages over North Carolina in terms of climate, recreational opportunities, and other quality of life considerations. If the state of Florida had acted to upgrade two or three state universities after World War II, all the ingredients would have been assembled for the kind of development the Research Triangle has experienced in terms of the growth of research activities.

Similarly, Texas had an extraordinary period of economic growth after World War II, and its oil wealth should have made financing universities and research facilities a simple matter. The state also had a disproportionate share of political power at the federal level, which could have been exerted to locate federal research activities in Texas (as evidenced by NASA's movement to Houston). The state university system in Texas was one of the richest in the nation, and the University of Texas at Austin seemingly had little excuse for not becoming a first-rate university. Both Florida and Texas have continued to experience extraordinary economic growth over the last thirty years, but neither was able to capitalize on its potential for becoming a center for research and development on the scale of Silicon Valley and Boston's Route 128.

In the early 1950s the Research Triangle Park had the necessary prerequisites for success, but several factors stand out as contributing to its subsequent development. First, in the initial development phase in the 1950s and early 1960s, the Research Triangle had the benefit of the three nearby universities, which contributed to the human capital resources of the area and the quality of life for incoming professionals. More importantly, the universities were heavily involved in the initial organization of the Research Triangle. This was a conscious policy decision on the part of organizing boards and committees of the Research Triangle Foundation and the Research Triangle Committee. Professor George Simpson from the University of North Carolina at Chapel Hill served as Executive Director of the Research Triangle Committee and involved faculty in industrial promotion trips and organizational arrangements.[3] The working relationship with the university faculties was also cultivated by the creation of interlocking directorates for the Research Triangle Foundation (RTF) and the Research Triangle Institute (RTI), with university faculty and administrators serving on both. University faculty have served as consultants on RTI projects since the beginning of the Institute. Adjunct professorships at the universities can be arranged relatively easily for professionals in RTI and the park's other establishments (this is an attractive tool for recruiting high-caliber scientists); today there are about four hundred such positions. RTI staff and employees of corporate and government facilities in the park can likewise enroll in courses at the universities.

Second, key to the success of the early financing arrangements for the land purchases was the extensive involvement of a relative handful of the state's bankers, industrialists, and politicians—including the governor and state treasurer—and the use of the institutional vehicle of the nonprofit corporation. The nonprofit nature of the Research Triangle Foundation enabled RTF to promote itself and the Research Triangle as a quasi-public industrial development organization without seeming too self-serving. For example, George Herbert, the Executive Director of the Research Triangle Institute, received state appropriations for equipment for RTI laboratories in 1959; the General Assembly made an unconditional grant to RTF for $750,000 in 1965 in order to attract the National Institute of Environmental Health Sciences (NIEHS); and, as discussed below, the General Assembly has allocated over $40 million to the nonprofit Microelectronics Center of North Carolina. It would have been impossible to make these state appropriations to private corporations.

The nonprofit status of the Research Triangle Foundation also meant that private donations to it were tax deductible and that the corporation did not have to pay taxes on its profits. Profits on land sales by RTF have been (and will be) very large. This has given RTF the flexibility to make special arrangements for and offers to organizations it really wants in the park and also to return assets to the Triangle universities in the form of both land and monies. In three cases the Research Triangle Foundation has donated land for a facility. The 26-acre site for the Forestry Sciences Laboratory of the United States Forest Service was provided to the federal government at no charge. The $750,000 unconditional grant from the General Assembly in 1965 enabled the foundation to offer a free site to NIEHS. RTF also gave a 120-acre site to the Triangle Universities Center for Advanced Studies, Inc. (TUCASI) in 1977, which enabled TUCASI to provide a 15-acre site and building for the National Humanities Center at a nominal rate. In 1981 TUCASI leased a 36-acre parcel from its 120-acre tract to the Microelectronics Center of North Carolina.

For RTI, nonprofit status offered a slight competitive edge over for-profit contract research firms and also a way to generate funds internally to finance growth. Perhaps most importantly, the vehicle of a nonprofit corporation enabled the founders of the Research Triangle to obtain much of the financing from public funds (directly and indirectly), while at the same time keeping control of the development out of the hands of the state and local government. There is little direct public oversight of the development decisions taken by the Research Triangle Foundation or the Research Triangle Institute.

Third, the Research Triangle benefited from the pleasant climate, the relative accessibility of the mountains and the coast, and the "liberal image" of North Carolina. Although the climate and natural amenities are not perhaps as dramatic as those of Florida, North Carolina was able to capitalize on the large-scale movement of industry and population to the Sunbelt. Evidence of this today is found in the growing retirement community in Chapel Hill, and the high rating given to North Carolina cities in quality of living studies.[4] The liberal image of North Carolina was especially important in the 1950s and 1960s during periods of racial tension in the South.[5] The school segregation crises in other southern states made it very difficult to attract professionals from northern universities, particularly those with school-age children. Governors Luther Hodges and Terry Sanford were able to exploit this advantage effectively in their industrial recruitment efforts.

Cooperation between the public and the private sectors in the establishment of the Research Triangle Park, along with the active involvement of the university hierarchy, was one manifestation of this "liberal image" and was itself impressive to industrial prospects. The state's political and financial elites effectively created an atmosphere in which things were being accomplished to promote the Research Triangle. Although the development of the Research Triangle was controlled and organized with very little citizen or public involvement in the decision making, all the available evidence suggests that this was being done with the public interest in mind, and without corruption. In all the press reviews and various accounts of the development of Research Triangle Park, there are no hints of financial mismanagement. Corporate leaders probably felt very comfortable working within this centralized, closed decision-making environment and appreciated the relative absence of politicians seeking private gain. Establishing a facility within the Research Triangle involved a particularly easy set of negotiations with the Research Triangle Foundation and the state government. Land prices were known in advance, and it was not necessary to involve brokers or make contributions to political candidates to strike a satisfactory deal.

The success of the Research Triangle was assured in the mid-1960s with the arrival of the "anchor" institutions: IBM and several large federal government facilities. The sale of land to IBM enabled RTF to retire the mortgage on its landholdings. IBM was courted by both Governor Hodges and Governor Sanford. Political influence seems to have been important in decisions to locate large federal agencies such as a division of the Environmental Protection Agency and NIEHS in the park. North Carolina's success in convincing federal agencies to locate facilities in the Research Triangle was in part due to Hodges's connections in the federal government resulting from his tenure as Secretary of Commerce during the Kennedy administration and to Sanford's collection of some political debts within the Democratic party. This ability to exercise political influence at the national level for the benefit of RTP can only be ascribed to good luck. These federal facilities could easily have been located elsewhere in the country. The presence of these major institutions provided the Research Triangle with both a corporate and a government "stamp of approval" and generated a good deal of very favorable national press coverage. This visibility aided greatly in the industrial recruitment process, inviting inquiries about the park and opening doors for North Carolina promoters. If the Research Triangle Park was good enough for IBM, perhaps it was worth taking seriously.

The role of the Research Triangle Institute should be stressed as an important factor in the successful development of the Research Triangle at every stage of its growth. As originally conceived, this research institute was to stimulate future growth by providing contract services to corporate and government clients and giving easy access to university faculty resources. This has in fact happened. RTI has experienced phenomenal growth over the last twenty years and has certainly strengthened the entire Research Triangle Park community. One of the lessons from the Research Triangle experience appears to be the importance of a successful, nonprofit research institute serving as an anchor institution.

Two final factors worth mentioning in the success of the Research Triangle Park were the ready availability of a large tract of land near the three universities and the one-day air service from Raleigh-Durham Airport to Washington, D.C., and to corporate headquarters in northeastern states. Though clearly not a sufficient condition for a successful research park, the size of the acreage reserved for RTP clearly indicated to corporate leaders that the site had the potential to become a very large-scale development. It was attractive from the perspective that they might be buying into something that could turn out to be very important. This remains a significant selling point today, and only one-half of the park is filled.

These factors that contributed to the growth of the Research Triangle Park are still largely in place, and the state's current policy initiatives to attract microelectronics firms are assumed to complement them and build upon past successes. Yet the character of past development in the Research Triangle Park has caused some speculation about whether the Triangle area might become a center of the kind of economic activity witnessed in Silicon Valley or around Route 128 in Boston. From its inception RTP has indeed been compared with the development around the Stanford Research Institute and Route 128. However, the private-sector establishments in the Research Triangle Park have been essentially branch research and development operations of large, "blue chip" corporations. Many of these (for example, IBM, DuPont, General Electric, Data General, Burroughs Wellcome, and Union Carbide) are firms based on the East Coast that have moved a portion of their research and development operations south to North Carolina. Corporate and governmental research personnel have interacted in RTI projects and seminars and in university and social affairs, but the Research Triangle has not witnessed the extensive entrepreneurial activity characteristic of Silicon Valley or Route 128.

According to George Herbert, president of RTI, no individuals have left the institute to start their own firms, either in the Triangle or elsewhere. Nor has the Triangle experienced the intense competition for highly skilled engineers and scientists that is common in Silicon Valley or around Route 128. In fact, the relative lack of competition for research talent, coupled with the availability of recent graduates from the universities, makes the local labor market one of the important attractions of the Research Triangle for major corporations, which seek to avoid rapid turnover among highly trained technical personnel.

Though they share branch plant status with other important industries in North Carolina, the branch research and development operations characteristic of the Research Triangle Park are only remotely linked with the rest of the state's economy. As discussed by Goldstein and Malizia (this volume), most of the facilities have few direct interactions with suppliers or final markets within the state. There are a few notable exceptions: Burroughs Wellcome has established a large manufacturing facility in Greenville; Glaxo, Inc., is building a manufacturing facility in Zebulon; Data General has manufacturing plants in Wake and Johnston counties; and IBM has spawned a variety of related activities. Nevertheless, one of the striking aspects of the establishments in RTP is their lack of integration with the rest of the North Carolina economy. If this proves true for microelectronics facilities that locate in North Carolina, it will have important implications for the achievement of the stated objectives of Governor Hunt's microelectronics policy.

The Establishment of the Microelectronics Center of North Carolina

To launch his effort to attract microelectronics firms to North Carolina, Governor Hunt adopted an organizational strategy almost identical to that used by Governor Hodges fifteen years before in the formulation of the institutional arrangements for Research Triangle Park. In the spring of 1980 Hunt formed a blue-ribbon committee of the heads of Duke University, the University of North Carolina at Chapel Hill, and the state's community college system, and charged them with the task of exploring the feasibility of launching the state into a microelectronics boom similar to that of California's Silicon Valley. The committee reported back to the governor in about six weeks, supporting the idea of microelectronics development in North Carolina. The committee apparently undertook no analytical or empirical investigations, nor did they have any significant amount of staff assistance, but they suggested that the way to accomplish the governor's objective was for the state of North Carolina to form a nonprofit corporation—the Microelectronics Center of North Carolina (MCNC)—that would sponsor research and development efforts and train students in microelectronics technology and applications. The establishment of such a coordinating institution was apparently the only policy explored in any depth for achieving the governor's objectives for growth of microelectronics in the state. This may have been due in part to the tone of urgency in the committee's report to the Governor. Press accounts of the committee's report have quoted the following conclusion: "It is almost too late to enter the microelectronics race: next year will be too late."

A nonprofit corporation, MCNC was to be formed and governed largely by five universities in the state: Duke, North Carolina A&T State University, North Carolina State University, the University of North Carolina at Chapel Hill, and the University of North Carolina at Charlotte. The model for the creation and organization of the center was thus essentially the Research Triangle Institute. In fact, MCNC could easily have been originated as a laboratory or branch of RTI, except that it would not have had the political visibility required.

The major difference between the establishment of MCNC and previous development efforts in the park was that Governor Hunt sought the majority of funding in the form of direct state appropriations from the General Assembly, rather than applying to private donors or slowly building the facility from internally generated contract revenues from RTI. Corporate membership in MCNC was to provide modest funds during the early years. In June 1980, Hunt committed almost $1 million from the state's contingency and emergency fund to found the center. The expenditure of these funds did not require legislative approval. In August General Electric announced that it had chosen North Carolina as the site for its new facility and that a key factor in the decision had been the establishment of MCNC.

An article in 1981 reported how General Electric had proceeded in its search for a location for its microelectronics facility. Beginning in 1979, it had first narrowed the possibilities to thirty locations. Then, as one of the managers of the new plant explained it,

> We scaled it down to six [locations] on the basis of technical labor force in the area and university environment; you see, a lot of folks like to continue their education so you want the academic.
>
> We also wanted a certain quality of life. There was Nashua, N.H. (close to Boston); Manassas, Va., outside Washington, D.C.; Orlando, Fla.; Albuquerque; Colorado Springs; and the Research Triangle. We went to each and this place won hands down.
>
> There's an ongoing semiconductor ethic here that goes back to the early sixties, and it turns out that this location is central to G.E.'s East Coast locations. We don't have much west of Milwaukee.[6]

General Electric also suggested that the three major universities in the Triangle cooperate to work in silicon technology. It offered some money to initiate this research, but wanted additional government and university funding. Governor Hunt was to obtain these funds from the General Assembly in the appropriation for the Microelectronics Center.

In November 1980 Hunt made a much-publicized trip to Silicon Valley, consistent with the long tradition of southern governors making industrial recruitment forays—except that Hunt headed to the West Coast rather than New England or the Midwest.[7] His message to microelectronics firms was that North Carolina had plenty of industrial development sites, cheap land, cheap housing, a relaxed quality of life, and a new microelectronics research center—all within a short distance of three major universities.

By using the state's contingency and emergency funds to start MCNC, Hunt had avoided the scrutiny of the legislative budgetary process. When the General Assembly met in January 1981, the legislators were in a sense faced with a fait accompli. The governor's budget for the 1981–1983 biennium included a $24.4 million appropriation for MCNC; he argued that the state could not turn back now, that $1

Figure 1. Organization of the Microelectronics Center of North Carolina

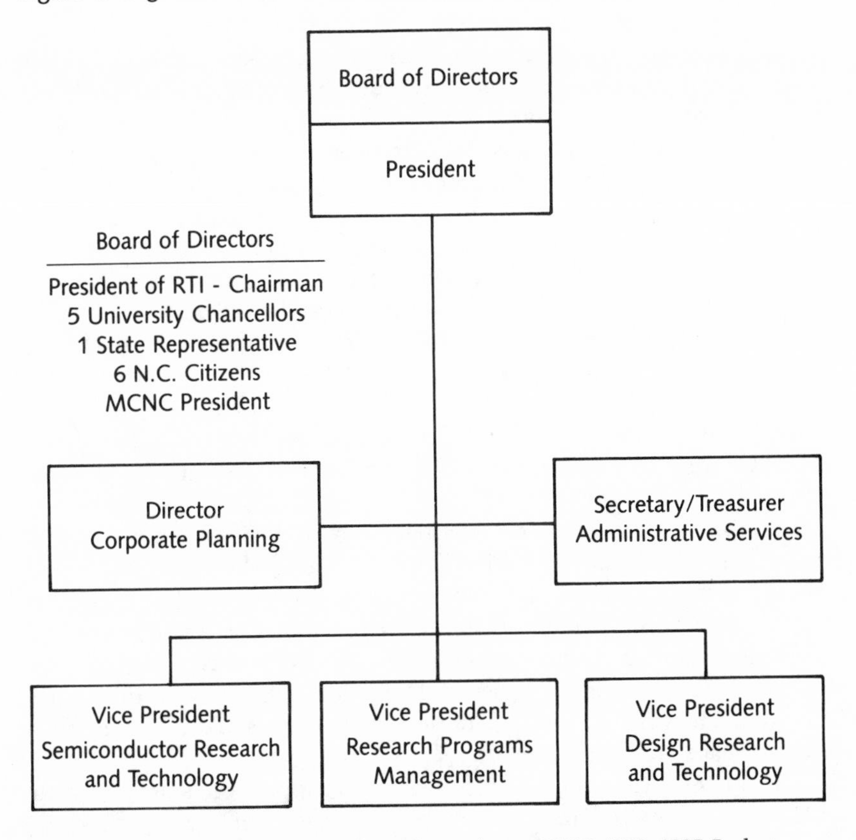

Source: Microelectronics Center of North Carolina, *MCNC 1985–1987 Budget* (Research Triangle Park, 1984), 7.

million had already been spent, that General Electric was already committed to moving to North Carolina, and that he had personally been out to California and already announced the creation of the Microelectronics Center. Hunt had put his political reputation on the line. Despite a tight fiscal environment and budget-cutting efforts in almost all other areas, the appropriation for MCNC never even had a public hearing. It passed in May 1981.

MCNC was established as a nonprofit corporation governed by a board of directors consisting of the president of the Research Triangle Institute, the chancellors of the five universities involved, one representative of state government, six North Carolina citizens appointed by the governor, and the president of MCNC. Figure 1 outlines the basic organization of MCNC. Unlike MCC in Austin, Texas, MCNC is primarily a public sector initiative intimately tied to the five participating universities. Industry involvement is necessary to MCNC's ultimate success and is being actively sought, but unlike MCC, MCNC was not formed as a private consortium of participating companies. Table 2 lists the active industrial participants in MCNC's programs as of 1984. The center offers semiconductor manufacturers, systems companies, and other related microelectronics firms an opportunity for broad involvement in its activities through its industrial affiliates program. An initial three-year membership costs $750,000 (or an equivalent contribution) and allows staff from the affiliate firm to participate in MCNC research programs and work with its facilities. Ownership of inventions or products arising from MCNC's research efforts resides with MCNC, but affiliates are given an advantage in acquiring rights to these on a preferred royalty basis. The first major product from MCNC's research efforts was VIVID, a software system for custom VLSI (very large scale integration) design.

Research and development is under way at MCNC in three main areas: (1) semiconductor materials, devices, and fabrication processes, (2) computer science and computer-aided design, and (3) integrated circuit design to support advanced microelectronics applications. To support its research and education efforts, MCNC is committed to providing state-of-the-art design and fabrication facilities for integrated circuits. The MCNC building presently contains a 10,000-square-foot "clean space" fabrication facility.

One of MCNC's primary objectives is to support the state of North Carolina's industrial recruitment of microelectronics companies. To do this, it assists the Department of Commerce with industrial prospects and visitors and generally seeks to create a national and international awareness of North Carolina's growing potential in the field of microelectronics. Since its inception, MCNC has achieved a high national and international visibility through participation of its personnel in congressional hearings, national conferences, and various commissions.

To date, MCNC has focused much of its energy on developing means to promote the training of engineers and scientists and to create a skilled labor force for the microelectronics industry. One of its brochures presents the rationale as follows:

> The best incentive North Carolina can offer the microelectronics industry to invest in new plants and jobs here is a state commitment to help train and educate men and women in specialities relating to microelectronics.

Table 2. Industrial Involvement with MCNC

Active Participant	Formal Invitation/Negotiation
Aeronca Electronics	AMD
AIRCO Industrial Gases	Analog Devices
Allied Chemical	Applied Materials
Bruce/BTU	Burroughs
Digital Equipment Corporation	Data General
Gaertner Scientific Corporation	General Motors
General Electric	Hewlett Packard
GCA/IC Systems	Intel
HHB Softron	ITT
Hybrid Technology Group	Motorola
IBM Corporation	National Semiconductor
Lexidata Corporation	
Metheus	
Monsanto	
Mosaic	
Millipore	
Northern Telecom	
Norton	
Philips Electronics Instruments	
The Plessey Company	
Semifab	
Shipley	
Siecor FiberLAN	
Varian	
Vashaw Scientific/E. Leitz	
Vectrix Corporation	
Zeiss/Eastern Microscope	

Source: MCNC 1985-1987 Budget, 26 October 1984.

> Just as textiles and other industries have depended upon higher education for progress, so must the microelectronics industry. Because microelectronics is a relatively new, but rapidly growing industry, the universities throughout the nation are not keeping up with its demand for people. Developing the facilities and staff to provide the necessary education requires a major financial investment. States willing to make this investment will reap the benefits of jobs for their young men and women by attracting the industry to this trained talent.[8]

MCNC's strategy for accomplishing this objective is to strengthen university research and education programs in terms of staff and equipment.

An obvious question to ask is why the state created a new institution to support university training, rather than allocating more money to university budgets for computer science and electrical engineering programs. There are several explanations. First, MCNC is a much more visible and politically useful symbol for Governor Hunt's initiative than a beefed-up computer science department. Aside from that, it would probably have been impossible to obtain either the initial $24.4 million state appropriation or a subsequent $18 million (in 1983–84) if the funds were to have been included in the state's general university budget. Second, MCNC provides an institutional arrangement by which university faculty in computer sciences and electrical engineering can be paid more than would be possible within the university pay structure. Faculty recruited in these disciplines can be offered a position or affiliation with MCNC that would supplement their university salaries. This increased flexibility in salary negotiations is crucial if the Triangle universities are going to be able to attract and keep top-caliber people in the microelectronics field. Third, the laboratory facilities and equipment being purchased by MCNC for joint use by the associated universities are simply too expensive to replicate on each campus; its central location likewise affords easy access for most participating universities.

Of MCNC's initial budget allocation of $24.4 million from the state, about $10.5 million was designated for the construction of a new 100,000-square-foot building in the Research Triangle Park to house offices and laboratories, and $8.6 million was designated for the purchase of equipment. Another large budget item was the $6.5 million telecommunications system that links campuses and private corporations around the state with a real-time, interactive system for broadcasting lectures in computer science, electrical engineering, and other sciences. This telecommunications system attempts to make the most effective use of faculty resources, in effect leveraging faculty time and helping to minimize the fragmentation of facilities and faculty among the five universities. The telecommunications system also serves as an impressive high-technology showpiece for visitors.

During the early phases of Governor Hunt's microelectronics initiative, there was little public discussion of the continuing public support that MCNC would require. In his 1983–84 budget Hunt requested an additional $18 million appropriation for MCNC, to be used for continued expansion of its programs. MCNC's budget request for fiscal years 1986 and 1987 totals to about $23 million in state funds. Estimates of annual operating expenses after the center is fully established are on the order of $20 million, two-thirds of which would be funded by direct state appropriations.

MCNC's budget is thus relatively modest when compared with its economic development objectives and its political visibility. Governor Hunt needed to create high public expectations for his microelectronics policy in order to sell it in a time of budgetary stringency and ideological pressure against public sector involvement in the free market. Even if North Carolina's effort does not live up to these expectations, the state budget allocations to MCNC may still prove to be money well spent.

Debate over Microelectronics Policy

Although MCNC's first appropriation breezed through the legislature, Governor Hunt's new microelectronics policy did touch off a lively public debate. The initial enthusiasm for microelectronics was extreme. Hunt himself proclaimed that "the microelectronics industry is our chance—perhaps the only chance that will come in our lifetime—to make a dramatic breakthrough in elevating the wages and per capita income of the people of this state." Howard Lee, Secretary of the Department of Natural Resources and Community Development in 1981, said: "Microelectronics could be the goose that lays the golden egg."[9] Media interest in the subject has been extensive, and the issues involved have drawn a wide assortment of critical comments, pro and con.

The governor's microelectronics policy was subjected to close scrutiny by the media in part because Hunt had decided that the only practical way to obtain the funding on the scale contemplated was from the state budget. That policy may indeed have been a good one, but the governor was not adequately prepared to defend it in terms of a detailed understanding of the structure and organization of the microelectronics industry. Part of the problem was that in their public statements Hunt and his staff never clearly differentiated between the widespread use of computers in society and the production process involved in making silicon chips. Hunt's political instincts, however, were good, and he clearly struck a political nerve. High-technology economic development is a powerful political symbol in a low-wage state, and good politics just about everywhere. When Walter Mondale came to North Carolina during a one-day campaign swing in 1982, one of the handful of places he visited was the temporary headquarters of MCNC. In November 1982, Adlai Stevenson III appeared on David Brinkley's Sunday morning national news program during his race for the governorship of Illinois. Stevenson was promoting traditional Democratic issues like economic development and jobs, and when Brinkley asked him what he would actually do if he became Governor of Illinois, Stevenson cited the "successful microelectronics program of Jim Hunt in North Carolina."

Much of the public discussion of Governor Hunt's microelectronics policy can be summarized in five related questions:

1. Are high technology or microelectronics firms likely to move to North Carolina?
2. Would the expansion of high technology industry—particularly microelectronics—be a "good thing"?
3. Assuming the growth of the microelectronics industry is desirable, does the state need to invest its resources to promote or accelerate this growth?
4. Assuming the state should target resources for the support of the microelectronics industry, what is the appropriate state policy to achieve this objective?
5. If the microelectronics industry does locate in North Carolina (whether or not as a result of state policy), what should the state and local government do to anticipate and plan for the consequences of such growth?

Chapters 6 and 7 of this volume provide several insights into the first question: whether high-technology firms are likely to locate in North Carolina. In chapter 6, Hekman and Greenstein illustrate that the shift in manufacturing employment to the South from 1950 to the present is broad-based and is occurring in such industries as plastics, fabricated metals, and electronics, not simply in the traditional mainstays of the North Carolina economy such as furniture and textiles (although in absolute terms North Carolina's economy is not very diversified). On the basis of a survey of recent manufacturing site selection decisions in Virginia, North Carolina, and South Carolina, they conclude that the most important business location factors for the southeast are lower production costs and the state and local industrial climate. Using the computer industry as an example, they also show that the South Atlantic region is primarily a center for manufacturing, as opposed to research and development. Their profile of high-technology facilities in North Carolina shows that these tend to be branch plants of corporations with several other branches that have headquarters in the Northeast or Midwest. In chapter 7 Malizia argues, however, that although lower production costs and state and local business climate may be important incentives for locating manufacturing in the Southeast in general, community quality of life factors, availability of skilled labor, the quality of the education system, and local infrastructure are more important attractions for high-technology manufacturers—and specifically semiconductor manufacturers—than for manufacturing establishments on average.

On the basis of the experience of the last twenty years, particularly the success of the Research Triangle Park and the broad-based movement of manufacturing to the Southeast, it seems highly likely that the Raleigh–Durham–Chapel Hill metropolitan area will continue to be attractive to many manufacturing and service-related sectors, including certain types of high-technology industry. Industrial location studies confirm that this is a very desirable place for relocation.[10] Moreover, recent quality of life studies rate Raleigh–Durham as a very attractive place to live. *Places Rated Almanac* ranked the Raleigh–Durham metropolitan area ninth out of 277 places in the United States.[11]

As discussed by Goldstein and Malizia in chapter 9, most industry observers agree that the Research Triangle area of North Carolina will receive some of the microelectronics industry's expansion outside California, but not necessarily the research and development segment of the industry. The most likely segments of the microelectronics industry to locate in North Carolina are probably wafer fabrication and branch operations of large, vertically integrated manufacturers of silicon chips who have the ability to specialize and diversify their research and development operations. Branch research and development operations typical of past development in the Research Triangle Park are another likely growth area in the high-technology sector. It would still be risky for merchant firms to relocate their research and development or corporate headquarters away from the Silicon Valley or Route 128, or for small enterprises to cut themselves off from the centers of innovation and creativity in the industry.

If such patterns do develop in the type of microelectronics and other high-technology facilities that locate in North Carolina, there will be important implications with regard to the second question: whether the development of the microelectronics industry in North Carolina would be a "good thing." Sampson points out in chapter 10 that if a typical wafer fabrication facility were established in North Carolina, the average wage paid would be about the current state average. However, because wafer fabrication facilities typically employ both high-wage professionals and low-wage production workers, this average reflects a balance of extremes rather than a realistic working wage for the majority of employees. The beneficiaries of such development would be largely the unemployed, individuals already in low-wage jobs who manage to find a low-skill job in a wafer fabrication facility, and the professionals at the upper end of the wage scale. Even so, a substantial number of positions in both categories would probably be filled by personnel coming from outside the Triangle area and North Carolina, because the Raleigh–Durham metropolitan area already has one of the lowest unemployment rates in the United States.[12] Moreover, the branch research and development operations characteristic of Research Triangle Park typically recruit new personnel nationally and also transfer significant numbers of employees from other corporate operations.

Sampson, Bourgeois, and Stein (chapter 2) and Robinson (chapter 3) observe that the microelectronics industry is maturing and becoming more capital-intensive. Barriers to entry are rising, and the microelectronics industry per se will not provide a wide range of opportunities for entrepreneurs or local businesses in North Carolina. Moreover, microelectronics is not likely to be a very prominent or very attractive employer. The major criterion for employment in the fabrication and assembly segments of the microelectronics industry is the ability to maintain a high level of attention to detail throughout the working day, not any specific computer-related technical skill.

Public concern over wage levels and microelectronics economic development policy, however, goes beyond simply a discussion of the wages paid to workers and the quality of work in the semiconductor industry per se. There is widespread concern that the "microelectronics revolution" will spur automation, which will in turn lower skill requirements and wages throughout the economy. Adler (chapter 4) critically examines this argument by analyzing the impact of automation on the nature of work in low-skill jobs in Demand Deposit Accounting in the French banking industry. He looks in depth at the changes in twelve categories of "work contributions" and concludes that technological change resulted in both increases and decreases in certain kinds of skill requirements, even among this segment of workers. Adler argues that the effects of automation are both varied and complex, and that a management policy that attempts to use new technology to downgrade skill levels will prove myopic in the context of the international competitive pressures facing many U.S. industries.

Although it seems unlikely that Governor Hunt's microelectronics initiative will "make a breakthrough in elevating the wages and per capita income of the people of North Carolina," that does not mean that the policy is a bad idea, or that average wages will not increase in a few local labor markets—but rather that as stated, the policy's objectives are unrealistic. A new microelectronics facility that pays average wages may well be a welcome addition to a local economy with significant unemployment or to one faced with the threat of plant closings in older industries. The governor's objectives of both higher wages and balanced growth would best be served if microelectronics facilities were located throughout the state, not just in the Triangle area and the central Piedmont. But judging from the conclusions of Malizia's research (chapter 7), this seems unlikely given the probable pull of the amenities of the Triangle area for professionals in the industry.

Any branch research and development or wafer fabrication facilities that locate in the Raleigh–Durham metropolitan area or surrounding counties would also benefit some existing residents in terms of increased property values, higher overall wages in the local labor market, and increased economic activity. Such benefits of economic growth are not, however, evenly distributed among area residents, but rather accrue largely to property owners in terms of increased real estate values, to owners of local businesses, and to other groups that can capture the economic rents associated with economic growth. For some area residents the principal consequences of a "successful" microelectronics policy may well be increased congestion, air and water pollution, and other undesirable aspects of urban sprawl.

Reacting to negative publicity concerning the exorbitant property values and environmental degradation in Silicon Valley, some critics of Governor Hunt's microelectronics policy accepted the proposition that growth of the microelectronics industry is likely in North Carolina, but questioned whether the increased economic growth would result in an improved quality of life. Essentially a growth versus no growth argument, this criticism had little to do with microelectronics policy per se except that, as it had in California, such development could apparently happen very quickly. This point of view has little political strength in a low-income state like North Carolina, particularly given the absence of anyone more environmentally conscious than Hunt on the North Carolina political landscape. Wariness of rapid development has, however, a long tradition among southern intellectuals dating back to the Agrarians' defense of the southern way of life, and the local media have been very alert to the issue.[13]

Aside from growth versus no growth debates over the wisdom of Hunt's microelectronics initiative, there appears to be disagreement among the state's economic and political elites concerning its advisability. Consistent with Hekman and Greenstein's conclusions (chapter 6), Luebke, Peters, and Wilson (chapter 12) argue that economic development of North Carolina in the postwar period is resulting in a broader industrial base and that these changes have resulted in a split within the state's political and economic elites—between those associated with traditional industries such as furniture and textiles, and those associated with the newer industries such as fabricated metals, electronics, and chemicals. Examining the sources of campaign contributions in five statewide races during the 1970s, they illustrate that Jesse Helms, a conservative Republican in the national Senate, relied on the traditional low-wage industries in North Carolina for support, while Hunt, a Democrat, was supported in his gubernatorial campaign by newer industrial sectors. The authors contend that the debates over the Microelectronics Center of North Carolina should be seen within the context of political conflicts over the proper role of the state in economic development. Hunt and his supporters viewed MCNC as a vital part of their effort to modernize and diversify the North Carolina economy; Helms and other conservatives saw it as a bad case of state planning and favoritism for out-of-state corporations at the expense of North Carolina's traditional industries.

As public discussion developed during 1981 concerning some of the negative consequences of locating wafer fabrication facilities in North Carolina, Governor Hunt argued that wafer fabrication was not, in fact, the segment of the microelectronics industry that would be attracted to North Carolina; rather, the state would attract research and development operations that paid higher wages. If this were to occur, it would almost certainly happen predominantly in the Triangle metropolitan area because of the nearby universities and other quality of life considerations important to engineers and scientists. Another argument advanced by the governor's office in support of the microelectronics initiative was that the real benefits in terms of higher wages and job creation would result from spin-off activities from the microelectronics industry and from regional multiplier effects, not from direct employment in wafer fabrication facilities.

In chapter 9 Goldstein and Malizia provide several insights into this issue. Considering the question whether new semiconductor manufacturers locating in North Carolina would buy from and sell to particular North Carolina-based firms on a significant scale, they conclude that one cannot expect important or unique trade effects. Their review of industrial cluster analyses suggests that proximity of "backward linked" industries that supply materials to electronic components manufacturers is not a critical factor in the location of manufacturing facilities, probably in part because shipping costs for the required materials are low. Moreover, though intraregional effects from "forward linked" trade (with manufacturers of communications equipment, computers, scientific instruments, and other products that operate with electronic devices) may be more important, the documented physical proximity of these industries to microelectronics manufacturers is probably due primarily to other common interests, such as technically skilled labor pools, nearby research institutions, or a preferred milieu in which to live and work.

The issue of how much the recruitment of the microelectronics industry would encourage spin-offs and the creation of new businesses is considerably more uncertain and may be partially determined by the aggressiveness with which the state recruits small businesses and supports indigenous enterprises. If microelectronics development in North Carolina is similar to existing development in the Research Triangle Park, however, the spin-off effects and business start-ups are likely to be minimal. Branch research and development operations of microelectronics firms would probably not employ many experienced corporate executives, who are the type of individuals most likely to initiate successful new enterprises.

With regard to the third question—does the state of North Carolina need to invest public resources to promote or accelerate growth in high-technology industries?—it seems clear that General Electric and many other corporations have found the Research Triangle an attractive location on the basis of educational facilities, local labor market conditions, and quality of life considerations. State-subsidized research efforts and manpower training were probably not primary considerations in their decision. This relative unimportance of state government financial incentives is borne out in Hekman and Greenstein's conclusions (chapter 6) and in other business location studies.[14] A "good business climate" and a certain working milieu are, however, often of major importance; these encompass a host of factors, including union activity, regulatory practices, and a perception of the willingness of the state and local government to accommodate industry needs. General Electric would probably have located its microelectronics facility in the Research Triangle Park irrespective of the state's action on MCNC, but the establishment of MCNC was a highly visible symbol of the state's willingness to cooperate with and support microelectronics firms—and the larger the state appropriation to MCNC, the more willing the state would have appeared.

In chapter 8 Luger examines North Carolina's high-technology and other economic development policies with regard to the state's goals of higher wages, more jobs, greater job stability, balanced development, and fiscal soundness. He then compares North Carolina to other states, in terms of both overall economic development efforts and high-technology initiatives. Putting North Carolina's high-technology policies in the perspective of its stated overall economic development program and goals, he contends that a disproportionate amount of attention has been paid to the recruitment of high-technology industry in comparison to the retention and creation of resident high-technology and traditional businesses. He recommends that North Carolina take closer account of the cost-effectiveness of its economic development efforts and broaden its menu of policies. Specifically, he argues that the state should focus more on capital assistance through the use of industrial revenue bonds, umbrella bonds, guaranteed loans, and interest subsidies to businesses it seeks to attract, and less on manpower training.

In the first of his two contributions to this volume (chapter 3), Robinson examines the nature of the innovation process in the microelectronics industry and the industry's organizational and locational characteristics, and concludes that the different segments of the industry that have developed have varying policy implications for states and localities seeking to attract microelectronics firms. He describes the merchant firms, captive suppliers, and custom design houses, and points out that the proportion of industry output represented by customized chips from both captive and independent suppliers is growing rapidly. Improvements in computer-aided design are expected to make customized circuit designs easier to develop and more economical. As the demand for customization increases along with further developments in circuitry, continuing start-ups and growth can be expected to occur in this segment of the industry. Moreover, captive firms, service firms, and custom houses are less tied to existing concentrations of the industry such as Silicon Valley. Robinson argues that economic development planners should focus their efforts on meeting the needs of this sector of the microelectronics industry. His conclusions in chapter 3 imply that the focus of MCNC on strengthening university facilities and the development of highly skilled manpower may well be the most appropriate thing the state can do to attract this sector of the microelectronics industry.

Granted that the state has a positive role to play in promoting economic development, we arrive at the fourth question posed above: how to determine appropriate state policy or the mix of policies required for the development of the microelectronics industry. Current state policies in North Carolina have been heavily influenced by the existing model of public–private sector cooperation that has succeeded in the Research Triangle Park and is familiar to state and corporate officials. The vast majority of the state's resources for the promotion of the microelectronics industry have been and are being invested in MCNC in the hope that it will serve as a vehicle for such development.

In addition to the establishment of MCNC, North Carolina has initiated a training program for microelectronics technicians in the state community college system. The first courses, opened in the fall of 1982, were modeled on technical training courses offered in California. State-funded vocational training courses are a common industrial recruitment policy in many states, and North Carolina's microelectronics training program is building on a well-established institutional structure. Other economic development policy initiatives are arising in state government to supplement or complement Governor Hunt's microelectronics initiative. For example, the Biotechnology Center of North Carolina was established in Research Triangle Park in 1981, modeled after MCNC. The state's 1983 New Technology Jobs Bill included two programs designed to benefit small businesses: (1) the North Carolina Innovation Research Fund, monies to be used as equity financing for research projects conducted by small businesses, and (2) the Incubator Facility Program, which created the North Carolina Technological Development Authority to make grants to help local communities build structures to house small businesses during their early phases of development. None of these initiatives has any hope of obtaining state funding on the level of MCNC, and they were not intended to be policy alternatives to MCNC.

Robinson's second paper (chapter 5) provides several insights into another aspect of the appropriate role of state policy in achieving an objective of supporting the microelectronics industry: what do microelectronics firms expect to learn from university research programs that they are supporting and are asking state governments to support? To address this question, he examines the nature and importance of the links between the quality of local university research and training efforts and the prospects of a local microelectronics industry, and focuses on the relationship between miniaturization and innovation. University programs clearly increase the supply of skilled manpower—an important objective for high-technology firms—but university research also aids industrial managers in choosing between different approaches to innovation. Robinson speculates on the areas in which the industry could most benefit from university research and involvement.

In the last section of chapter 5 Robinson surveys the different institutional arrangements of various microelectronics research centers (MRCs) in American universities. He notes that one of the important functions of these MRCs is to increase communication between firms and university researchers. The possibility of this communication, the likelihood that university research will be useful to firms, and the effectiveness of graduate training are all enhanced when universities equip laboratories with modern testing and processing equipment similar to that used in industry. MCNC is similar to Stanford and MIT in its commitment to the acquisition of state-of-the-art integrated circuit processing capabilities. Robinson notes that although this strategy may prove effective in building links between firms and universities, it is expensive and is somewhat risky due to the possibility of technological obsolescence. He also points out that MCNC is noteworthy among MRCs in its attempt to establish itself without major corporate involvement or financial support. This lack of significant private sector involvement in the formation of MCNC, and in North Carolina's microelectronics initiative in general, is probably one of the reasons why the Microelectronics and Computer Technology Corporation (MCC) selected Austin, Texas, as a location rather than the Research Triangle—which is ironic given the strong public–private sector cooperation that occurred in the establishment of the Research Triangle Park itself.

In approaching the fifth question—the appropriate response of state and local governments to the expansion of the microelectronics industry in North Carolina—it must first of all be recognized that public debate on the issue has often failed to distinguish between projected impacts due uniquely to the microelectronics industry and those that would result from any industrial growth. In chapter 9 Goldstein and Malizia attempt to identify potential effects on the development of North Carolina's economy that would be specifically a result of growth in microelectronics industry. They conclude that in many respects growth in that industry would produce effects similar to those caused by the kind of broad-based industrial expansion already occurring in the state.

Two issues specifically related to microelectronics growth that have caused widespread concern are water use and the generation and disposal of hazardous wastes. Some critics of Governor Hunt's policy have charged that the industry is a heavy water user and would exacerbate local water shortages, that it generates large volumes of hazardous wastes, and that workers in existing wafer fabrication facilities have been exposed to a wide variety of exotic hazardous materials. These criticisms were defused relatively easily by the Hunt administration, largely on the grounds that technology was changing rapidly in the industry and that the dangers cited were related to obsolete production technologies that would not be used in North Carolina. Because it seems impossible for anyone to forecast technological innovation in the microelectronics industry, this proved an effective rebuttal of the critics, who were forced to attempt to determine whether likely technological changes would in fact improve the residuals discharged in the workplace and the environment.

Just as uncertainty about the environmental and health-related effects of microelectronics development has fragmented public criticism, it has also made regulation by state and local governments exceedingly difficult, as Runge points out in chapter 11. Reviewing the regulatory strategies North Carolina may employ to control potentially hazardous wastes from the microelectronics industry, he identifies a number of areas where additional information and regulatory reforms can limit the overall risks. He concludes that as a result of weak leadership and the erroneous view that industrial recruitment is in conflict with effective waste management, North Carolina continues to lag in the development of state and local regulations appropriate for high-technology development, concentrating too much of its attention on landfill siting.

A related problem is the lack of technical expertise in the public sector in North Carolina to deal with the complicated regulatory problems raised by an industry, such as microelectronics, that is based on a complex and rapidly changing technology. By the time the public sector could learn enough to formulate an appropriate response to a perceived problem, the policy might well be obsolete and counterproductive. North Carolina had a very difficult time finding someone qualified to head MCNC, much less finding policy analysts competent to address the emerging regulatory problems posed by the industry's growth.

Conclusion

North Carolina's microelectronics policy initiative is still in its formative stages, and it is too early to know whether the state will in fact become a major center of the industry—and if so, what the social and economic ramifications will be. The wisdom of Governor Hunt's efforts to attract the microelectronics industry to North Carolina is still a matter of public debate. Yet much of the art of political leadership is understanding how to use such symbols to make things happen. With relatively modest commitments of state resources, Hunt drew increasing attention from the national media to North Carolina's high-technology economic development efforts; and because North Carolina takes itself seriously, other people are beginning to as well. In a recently issued report on the location of high-technology industries in the United States, the Joint Economic Committee of Congress mentioned the Research Triangle along with Silicon Valley and Route 128 as a likely future center of these industries.[15] Objectively, the Research Triangle has nothing like the concentrations of high-technology industry that currently exist in Silicon Valley or on Route 128, yet the impression that it might become the next major high-technology complex is being carefully cultivated. This perceived potential is indeed vitally important to the success of North Carolina's economic development strategy for high technology, and from this perspective, the importance of the Microelectronics Center of North Carolina goes far beyond the direct consequences of its budgetary expenditures. It thus makes sense that MCNC spent over $10 million of its initial $24.4 million on an elegant building to serve as a showcase of its ideals, for users, visitors, and the media. By the winter of 1985 MCNC was already hosting approximately two thousand visitors a month.

The Microelectronics Center of North Carolina is thus already serving one of its primary purposes as a symbol of the state's commitment to become a major center of the microelectronics industry.

Notes

1. "Research Triangle Seeks High Technology Industry," *Science*, 12 November 1965.

2. See W. B. Hamilton, "The Research Triangle of North Carolina: A Study in Leadership for the Common Weal," *South Atlantic Quarterly* 65, no. 2 (Spring 1966); Louis R. Wilson, *The Research Triangle of North Carolina* (Chapel Hill: The Colonial Press, 1967).

3. Hamilton, "Research Triangle," 259.

4. Richard Boyer and David Savegean, *Places Rated Almanac* (Chicago: Rand-McNally, 1981); Center for Policy Studies, University of Texas at Dallas, "National Attitude and Awareness Study" (prepared for the North Texas Commission), December 1981.

5. James C. Cobb, *The Selling of the South: The Southern Crusade for Industrial Development, 1936–1980* (Baton Rouge: Louisiana State University Press, 1982), 147.

6. Roger Lopata, "Research Triangle: A Far Out Concept that Worked," *Iron Age*, 23 November 1982.

7. Cobb, *Selling of the South*, 64–95.

8. *Microelectronics Center of North Carolina* (Research Triangle Park: MCNC, n.d.).

9. Quoted in Michael Luger, "Promises and Policies: The Economic Hope of the Microelectronics Industry," *N.C. Insight* 4, no. 3 (1981), 27.

10. Alexander Grant & Co., *A Study of Manufacturing Business Climates in the 48 Continguous States of the U.S.*, Center for Policy Studies, National Attitudes and Awareness Study (Washington, D.C., 1980), 10–18.

11. Boyer and Savagean, *Places Rated Almanac*, 370.

12. Goldstein and Malizia, chapter 8, this volume; Luger, "Promises and Policy."

13. On the Agrarians see Cobb, *Selling of the South*, 1–2.
14. See R. W. Schmenner, *Making Business Location Decisions* (Englewood Cliffs, N.J.: Prentice-Hall, 1982), 45–54.
15. U.S. Congress, Joint Economic Committee, *Location of High Technology Firms and Regional Economic Development* (Washington, D.C., 1982).

Part One

National Perspective and Issues

An Overview of the Microelectronics Industry

Gregory B. Sampson,
Tom Bourgeois, and
James I. Stein

The electronics industry, defined by the Standard Industrial Classification system as SIC 36, encompasses a wide range of economic activity. The sector is composed of three primary subsectors, which manufacture very different products. The *electrical equipment* subsector includes power distribution and specialty transformers, switchgear, motors and generators, industrial controls, and telecommunications equipment. Analysts often take a broader view and include computers and peripherals, test and measuring instruments, and office equipment within this category. This entire grouping is described as "electronic capital goods," in contradistinction to the *electronic consumer goods* subsector. Electronic consumer goods include television receivers, auto radios, phonographs, hi-fi equipment, and audio and video recorders/players. The primary product of the subsector is color television.[1] The direction of change in consumer electronics is toward the "home entertainment center," that is, a *system* of components such as videotape recorders, video disks, personal computers, and other equipment centered around a sophisticated color television receiver. By and large the character of the microelectronics technology incorporated in these consumer final goods will determine whether their producers can compete in the marketplace on the basis of cost and quality. The third major subsector of SIC 36 is *electronic components*. One segment of the industry produces "passive" components, such as resistors, capacitors, wires, and cables. Semiconductor devices and electronic tubes comprise the other segment, "active" components (see figure 1).

This chapter concentrates on the current state and future prospects of the semiconductor, or microelectronics, industry (SIC 3674), which falls under the electronics subsector of electronic components. Currently, North Carolina has a very small semiconductor electronics industry. Discounting "captive" semiconductor suppliers, there were fewer than five hundred people employed in the North Carolina semi-

Figure 1. Component elements in the electronics sector of the United States, Japan, and Western Europe, 1977 (1974)

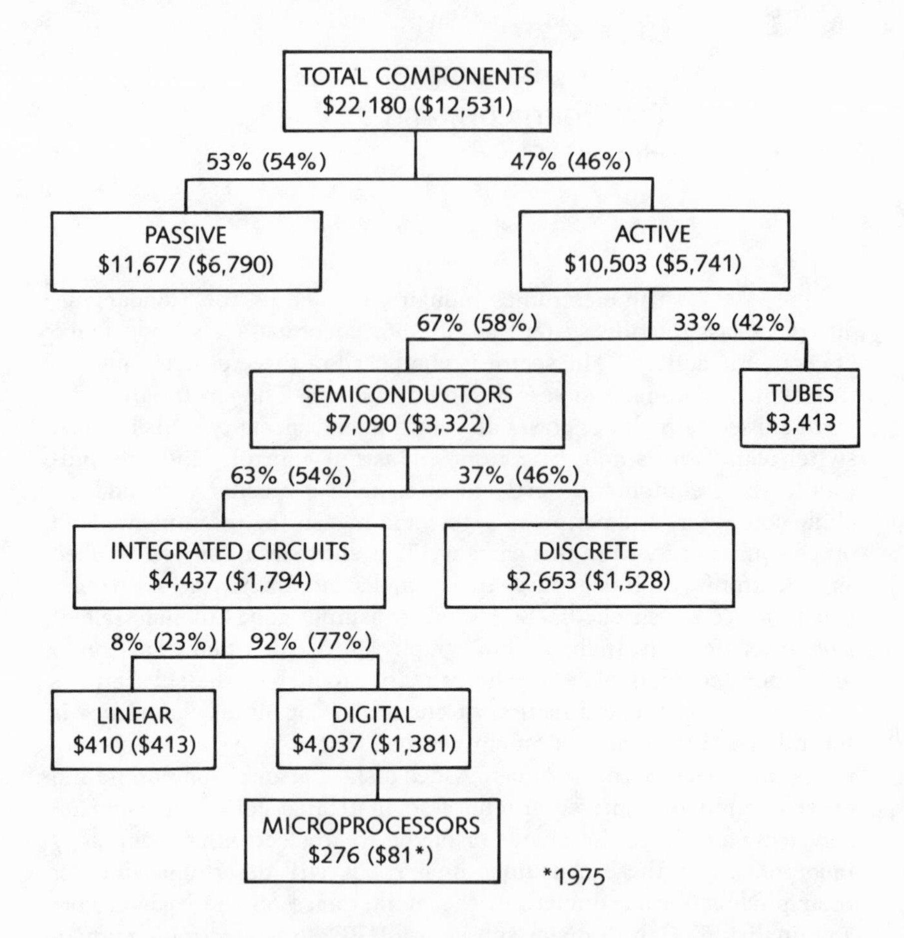

Source: Mick McLean, *Technical Change and Economic Policy: Sector Report, The Electronic Industry* (Paris: Organization for Economic Cooperation and Development, August 1980).

conductor industry in 1984. State officials have, however, been working quite avidly since 1981 to promote the idea that North Carolina is an up-and-coming southeastern "seedbed" for the microelectronics industry. The state has committed over $41 million (through fiscal year 1984–85) for the development of specialized curriculum and support programs and the construction of the physical plant for the Microelectronics Center of North Carolina (MCNC).[2] The state has also increased its financial commitments to university and community college programs that contribute to the ability of the state to attract and nurture microelectronics firms. These include funds for the departments of computer science at the University of North Carolina at Chapel Hill and North Carolina State University in Raleigh, high-technology training programs at five community colleges, and the acquisition of advanced microelectronics training equipment for community colleges.

Though it is still too early to tell whether these efforts will be successful, a December 1983 survey of American electronics firms by the Electronic Location File of Surrey, England, found that North Carolina had moved from tenth to fifth place in the firms' preference for industrial location and expansion sites. Though two other southeastern states, Florida and Georgia, were among the top ten, no southeastern state had a higher preference rating than North Carolina.[3]

The Economic Significance of the Microelectronics Industry

According to several recent estimates the electronics industry will rival the automobile, steel, oil, and chemical industries in sales by the end of the decade.[4] Electronics is currently a $100 billion industry in the United States,[5] and the worldwide market is forecast to be in the range of $600 billion to $800 billion by the late 1980s. One observer has suggested that "in the remainder of this century it will be the state of a nation's electronics industry that signifies whether it is a developed nation or not."[6]

Microelectronics is the driving force in the electronic revolution. Semiconductor technology and its evolution have been termed "the crude oil of the electronics industry."[7] Integrated circuits (ICs) are the centerpiece of microelectronics technology. The capacity to package 100,000 or more components on a single silicon chip, smaller than a postage stamp, has made information a low-cost item. Moreover, this capacity promises to bring about changes of such magnitude in industrial production, the structure and organization of commerce, and the character of consumer goods that analysts see it as the seed of a second industrial revolution.[8]

Traditional measures of economic importance undervalue the significance of the integrated circuit industry. On the basis of such standards as total sales, gross product originating, or employment growth, integrated circuits are not very distinguished. In fact, recent estimates

Table 1. Microprocessor Applications by Sector

Sector	Application	Examples
CONSUMER GOODS	Household domestic appliances	Washing machines, ovens, sewing machines
	Entertainment products	Television sets, video games, video recorders, hi-fi equipment
	Personal products	Cameras, calculators, watches, language translators
	Cars	Dashboard displays, engine control (ignition, exhaust), collision avoidance, braking systems, diagnostic systems, petrol pump control
COMPUTERS AND PERIPHERALS	Minicomputers	
	Memory equipment	Magnetic disk/drum control, semiconductor memories
	Input/output equipment	Keypunch systems, "intelligent" terminals, point-of-sale terminals, optical character readers, printers/displays, electronic funds transfer, modems
	Data transmission equipment	"Front-end" processors, multiplexors
TELECOMMUNICATIONS	Exchange equipment	Time-division multiplex transmission, telex switching systems
	Subscriber equipment	Viewdate terminals, teletypewriters
OFFICE EQUIPMENT	Data processing	Accounting machines, visible record computers
	Word processing	Word processors, copiers, facsimile
	Audio equipment	Telephone answering machines, dictation systems

Source: Mick McLean, Technical Change and Economic Policy: Sector Report, The Electronic Industry (Paris: Organization for Economic Cooperation and Development, August 1980).

show that integrated circuits comprised 23 percent of the components subsector but only 10.8 percent of electronics as a whole (see figure 1). The fundamental importance of the industry is more evident in the phenomenal pace of technological change, the rapid incorporation of digital-based integrated circuits into those sectors where electromechanical functions had been prevalent, and the prospects for the extension of microelectronic technology across the whole spectrum of consumer goods, industrial process controls, and commercial and information-handling applications (see table 1). Although the manufacture of integrated circuits was only a $10 billion world industry in 1980, its sales are presently growing at an annual rate of 30 percent. Even at current levels the industry has provided essential inputs to more than $250 billion in final goods.

Historical Background

The semiconductor industry developed after the transistor was invented at Bell Laboratories in 1947 by William Shockley, Walter Brattain, and John Bardeen. Although it sponsored the basic research that produced the transistor, Bell Laboratories was more interested in consuming than producing its invention. It shared the technology widely through publications, symposia, and liberal patent licensing arrangements, partly in an effort to allow others to improve the technology of production, and partly to escape the antimonopoly lobby.[9]

During the early 1950s the semiconductor industry was dominated by the large valve or vacuum tube manufacturers, which were the first companies to adopt transistor technology. Although these larger, traditional valve companies maintained their leadership in innovations, they were unable to exploit these innovations in the production process; in the late 1950s and early 1960s new, smaller firms with high production capacity and low unit prices came to dominate the discrete semiconductor market.

As market leadership shifted from the large established firms to the new, small firms, the location of the industry changed. It was originally an East Coast industry, largely because this was where the large electronics firms were headquartered. The new firms that formed in the 1950s located in the East as well, around Boston and Long Island. Analysts have cited several reasons for this locational preference: (1) these areas could provide the ancillary industries required by the semiconductor industry; (2) they were the homes of major universities; (3) they were cultural centers, which made them attractive residential locales; and (4) they were the centers of venture capital markets.[10] The financial climate for new, small firms was good. The Small Business Investment Act of 1958 established an asymmetrical tax treatment of capital grains and losses from investment in small enterprises, allowing capital losses to be deducted from personal income. This made investment less risky and stimulated the supply of venture capital during the late 1950s and through the following decade.

In 1955 William Shockley left Bell Laboratories to form Shockley Semiconductor Laboratories in his hometown, Palo Alto, California. In 1957 eight individuals left Shockley's firm to form Fairchild Semiconductor (also in Santa Clara County, California) as a subsidiary of Fairchild Camera and Instrument Corporation of New York. Fairchild Semiconductor became the progenitor of forty-one semiconductor firms, most of which also located in Santa Clara County.

Aside from these historical factors, the same reasons cited above for the clustering of new firms around Boston and New York applied to the San Francisco Bay area in California. Some analysts have cited an additional factor: the huge markets for semiconductors generated by the defense and aerospace industry contractors and subcontractors, which were also clustered in California.[11] This local demand for semiconductors, coupled with the large supply of engineering and science Ph.D.s produced by Stanford University and the University of California at Berkeley, made it easier for smaller, young firms without access to nationwide marketing and recruitment networks to gain needed manpower and contracts.

The clustering of firms in what is now known as Silicon Valley, as well as in Boston and Phoenix, seems to have resulted from the rapidity of technological change in the industry and its consequent dependence on skilled manpower. Market share in the microelectronics industry has been and continues to be very much a matter of timing—being in the right place at the right time with the right price—and thus it has been crucial that firms have access to the most up-to-date information on products and process technologies. Proximity that allows for the *informal* exchange of information has been essential to the success of small firms in the industry.

From its inception the industry has been characterized by a high degree of interfirm mobility of personnel, particularly among scientists and engineers. Indeed, many observers believe that this mobility of highly skilled professionals has contributed significantly to the industry's rapid progress. The opportunity to earn very large salaries and assume stock ownership options, as well as a chance to see one's own ideas successfully carried through production and to the market, have created strong personal motivations for rapid innovation. In fact, many semiconductor industry professionals have been unable to adapt to employment within large, diversified firms. These companies cannot adjust to the high-turnover, high-salaried, California style of business of the semiconductor industry, and industry professionals are often unable to adapt to the corporate control and relative organizational inflexibility of large firms.[12]

Yet although small firms have set the pace in technological progress in semiconductor production, they are not the major source of technological innovation in the industry. Basic and applied research activity has been concentrated in universities central to the field—Stanford and MIT—and in the experimental laboratories of large organizations like Bell Laboratories, IBM, and Texas Instruments. Before 1968, Bell Laboratories produced 56 percent of the industry's major process innovations; by 1968, Fairchild Semiconductor, IBM, and Texas Instruments held 44 percent of all product patents.[13]

Despite this concentration of major innovations within a few firms, rapid diffusion of technology has prevented their monopolization. The industry has maintained a liberal patent licensing tradition, beginning with Bell Laboratories' sharing of transistor technology in 1952. The industry's early dependence on military contracts created another tradition called "second-sourcing." To ensure the availability of products, the military often required innovating firms to make an agreement with another manufacturer to produce the product to the same military specifications, so that the two firms' products were interchangeable. These two procedures, in conjunction with high levels of personnel mobility and outright pirating of innovations, allowed rapid diffusion of technology and easy entry of new firms into the industry in the 1950s, 1960s, and early 1970s.[14] Most recently, as research and development have become more costly, and as semiconductor technology has become more complex, the diffusion of innovation to new firms has slowed. The expense of second-sourcing is becoming prohibitive, particularly for microprocessors, and longer-term procurement agreements—nineteen to twenty-six weeks—have become more prevalent to stabilize supply.[15]

Industry Performance, Structure, and Sales

Performance in the semiconductor industry has been very impressive. Sales growth was 36 percent in 1979 and 26 percent in 1980. Table 2 illustrates the superior performance of this industry as compared with all industry across four standard measures of profitability and growth. The semiconductor industry also shows strong performance according to a Bureau of Economic Analysis assessment scheme that includes such measures as material input prices, research and development effort, change in new shares of capital stock, unit labor costs, labor productivity, and industrial structure.[16] Research and development was 5.8 percent of sales in 1978, three times the average for American industries as a whole; its ratio to profits was 102.3 percent. Capital expenditures by the seven leading semiconductor firms are presented in table 3. The turnover rate of capital stock in the industry is extremely high. It has been claimed that in any given year 75 percent of sales dollars come from products that did not exist five years before.[17] Industry officials believe that special depreciation treatment should be accorded this industry because much of its production equipment becomes obsolete after only two years—another sign of the ferocity of competition within the industry and the extraordinary rate of technological change that has occurred there.

Generally speaking, the microelectronics industry has two main segments: merchant firms and captive firms (see tables 4 and 5). Merchant firms sell their products on the open market. Captive firms offer no products on the open market and do not maintain an external sales force; they produce semiconductors for use in the products of a vertically integrated company. In Europe and Japan, semiconductor suppliers are heavily vertically integrated, but, particularly in Japan, this integration does not preclude selling both to the open market and to the internal market of the parent corporation.

The total value of production by U.S. captive firms is difficult to assess, because reliable production data simply are not available. Industry analysts have estimated that captive firms produced about one-third the amount of integrated circuits manufactured by merchant firms in 1980.[18] Captive firms are believed to be increasing their market share of total semiconductor production, and some analysts believe their influence on the open market has been greatly underestimated. A number of captive firms have been installing enough capacity to affect the open market significantly.[19]

The main reason for captive expansion appears to be the need for continuity of supply of semiconductor devices. During the 1974–75 recession and again during the 1981–82 recession semiconductor merchant firms cut back their capital expansion programs. When the economy recovered and demand accelerated, firms typically did not have the capacity to meet the increased demand. Backlogs lengthened, and shortages occurred. At the beginning of 1984 backlogs in orders were as high as nine months. According to some industry analysts, another reason for expansion in captive production is that many electronics companies increasingly feel they are backward unless they have some integrated circuit manufacturing capability.[20]

Competition in an International Market

The advanced industrial nations of the world share the perception that a competitive microelectronics industry will be inextricably tied to the future economic health and security of their countries. Early on the United States established a technological dominance in this area that has continued to the present (see table 6). In 1976, U.S. companies captured 65 percent of the world semiconductor market, the Japanese had 18 percent, and European companies 16 percent. In the most advanced products, large-scale integrated circuits, the United States held an even greater lead, with an 82 percent share; Japan had 10 percent and Europe 8 percent. Analysts have noted several reasons for the United States' early dominance in the industry. During the 1950s and 1960s, when unit costs were still relatively high, the military market in the United States enabled early indigenous producers to expand capacity. Foreign producers never had access to this large and secure market. In addition leading innovators in microelectronics have characteristically been small, new firms. In Europe the large, diversified electronics companies have controlled the European market from the start, and European experts have not broken away from these large firms to create new companies. Venture capital has been more difficult to obtain in Europe, and scientists who work for large enterprises or large government research laboratories have been accorded more prestige.[21] Finally, neither Europe nor Japan has had an internal market for microelectronics the size of the U.S. market, which remains the largest for semiconductors in the world. In Europe producers must serve fragmented markets with different design needs; this has prevented them from achieving economies of scale that would permit competition with U.S. or Japanese manufacturers.[22]

However, U.S. hegemony in this industry is diminishing. Recent estimates from a survey of the markets within the industrialized world (United States, Japan, and Western Europe) indicate that by 1980 the U.S. share of the microelectronics market had fallen to 45 percent, the Japanese share had increased to 30 percent, and the European share had

Table 2. Relative Performance of Selected Firms in the Microelectronics Industry

Company	PROFITABILITY			
	Return on Equity			Debt/ Equity Ratio
	5-Year Average	5-Year Rank	Latest 12 Months	
Teledyne	36.1%	1	27.7%	0.2
Schlumberger	32.6	2	36.3	0.2
Intel	30.5	3	30.2	0.0
EG&G	26.1	4	38.8	0.1
Lear Siegler	24.2	5	22.4	0.6
Raytheon	22.7	6	23.6	0.1
National Semiconductor	22.4	7	29.1	0.4
Avnet	20.6	8	21.5	0.2
Harris Corp.	19.8	9	19.7	0.4
Texas Instruments	19.4	10	21.4	0.0
Tektronix	19.3	11	18.8	0.3
Hewlett-Packard	18.7	12	21.4	0.0
GK Technologies	17.1	13	24.8	0.6
Perkin-Elmer	16.7	14	20.7	0.3
General Instrument	16.5	15	20.9	0.2
Motorola	15.8	16	17.3	0.3
Corning Glass Works	15.4	17	13.5	0.2
Beckman Instruments	15.0	18	14.9	0.3
North American Philips	14.6	19	12.8	0.2
General Tel & Electric	14.4	20	9.9	0.9
Ampex	14.4	21	16.4	0.5
Honeywell	13.4	22	13.2	0.2
American Tel & Tel	12.7	23	12.8	0.7
Bunker Ramo	12.1	24	15.0	0.4
Varian Associates	8.1	25	12.2	0.4
Industry medians	17.1		20.7	0.3
All industry medians	15.8		16.1	0.4

Source: "Annual Industry Survey," Forbes, 5 January 1981.

moved up to 25 percent. Specifically in the area of integrated circuits, the United States held 50 percent of the market, Japan 27.8 percent, and Europe 22.8 percent (see table 7).

In 1978 several European nations began government-sponsored ventures to establish future competitiveness in microelectronics. Over the next several years the French government projected the expenditure of $140 million on five companies for very large-scale integration (VLSI)

PROFITABILITY				GROWTH		
Return on Total Capital				Sales		Earnings/Share
Latest 12 Months	5-Year Rank	5-Year Average	Net Profit Margin	5-Year Average	5-Year Rank	5-Year Average
21.1%	3	23.4%	11.6%	11.0%	20	51.8%
30.0	1	29.8	18.8	23.6	4	35.3
29.2	2	29.7	11.7	45.5	1	39.5
26.6	5	21.0	4.2	25.1	3	26.6
14.4	13	14.2	4.6	13.2	15	34.0
27.4	4	21.6	5.6	14.6	14	24.2
20.1	10	16.8	5.4	36.1	2	26.9
19.8	8	17.4	5.7	15.2	13	16.3
13.2	12	15.1	6.1	15.3	12	31.9
21.1	6	18.6	5.3	17.6	9	18.0
15.4	9	16.9	8.4	23.5	5	26.7
20.6	7	18.1	7.7	23.3	6	22.7
14.4	18	10.4	5.0	15.4	11	15.6
17.4	11	15.5	6.7	21.7	7	18.0
17.4	16	12.6	7.5	10.9	21	36.8
14.4	14	14.0	5.7	12.9	16	16.0
11.9	15	12.9	7.6	8.1	23	16.8
12.3	17	12.5	6.5	16.5	10	24.1
9.1	19	10.1	2.9	20.1	8	15.2
5.3	25	7.0	4.6	11.7	18	7.6
12.1	22	8.8	5.2	5.0	25	D-P
10.0	20	10.1	4.6	8.6	22	14.7
7.1	23	7.2	12.1	11.6	19	9.2
11.3	21	9.4	5.2	7.0	24	80.5
10.5	24	7.1	3.6	12.3	17	25.6
14.4		14.2	5.7	15.2		24.2
11.0		11.1	5.0	14.3		13.9

research; Great Britain allocated $110 million for its programs, and West Germany $150 million.[23] In 1976 the Japanese government began a well-publicized $250 million joint effort with five electronics giants to move to the forefront of VLSI circuit technology. The United States ran its first deficit in integrated circuit trade with Japan in 1978, of $3.4 million; in 1980 the deficit was estimated to be $240 million. Japanese penetration of the U.S. market has continued to increase and

Table 3. Capital Expenditures in the Semiconductor Industry ($ millions)

	1978	1979	1980 (est.)
Texas Instruments	$116	$200	$265
Motorola	72	159	200
Intel	104	97	150
Mostek	19	40	130
National Semiconductor	58	88	125
Signetics	36	48	85
Advanced Micro Devices	23	49	70
TOTAL	$428	$681	$1,025
Capital spending growth		+59%	+51%
Sales growth		+36%	+26%
Profit margin		6.6%	6.2%

Source: "Microelectronics Survey," The Economist, 1 March 1980.

has become a major concern to U.S. producers. The volume of Japanese shipments of integrated circuits doubled from the second half of 1979 to the second half of 1980. By 1981 the Japanese had approximately 70 percent of the U.S. market for 64K RAMs. The Japanese government's efforts to promote the nation's microelectronics industry in world markets have proved to be highly successful. Perhaps the best measure of the effectiveness of Japan's protectionist policies is that the government has been able to prevent the establishment of U.S.-based manufacturing and sales subsidiaries, with the exception of a single Texas Instruments subsidiary in Japan. (In contrast, U.S. semiconductor firms were able to establish forty-six subsidiaries in Europe as early as 1974.) In general, the Japanese government has managed to erect and benefit from a whole series of non-tariff barriers—restrictions in government procurement, import licensing, technical product standards, financial controls, customs procedures, and documentation, among others—surrounding their domestic microelectronics industry.[24] As well, government agencies in Japan spread research and development findings throughout the industry, and Japanese firms have had rapid

Table 4. Top Ten U.S.-Based Integrated Circuit Manufacturers: Integrated Circuit Production Estimates ($ millions)

1980 Rank	Manufacturer	1979	80/79 %	1980	81/80 %	1981
1	Texas Instruments	$907	36	$1,230	4	$1,280
2	National Semiconductor	490	47	720	11	800
3	Motorola	510	34	685	8	704
4	Intel	460	34	615	13	695
5	Fairchild	365	23	450	6	475
6	Signetics	265	48	390	13	440
7	Mostek	220	66	365	8	395
8	AMD	207	40	287	17	335
9	RCA	165	22	200	5	210
10	Harris	120	54	185	14	210
TOTAL		$3,714	35	$5,127	9	$5,580

Source: William I. Strauss, ed., Status '81 (Scottsdale, Ariz.: Integrated Circuit Engineering, Inc., 1981),

access to new technology through liberal licensing agreements with U.S. firms. Moreover, the large diversified firms that control the industry in Japan have been quick to adopt new technologies.[25]

Overall, though, the U.S. balance of trade in electronic components has been quite favorable; it stood at $640 million in 1980. Nevertheless, the industry remains highly competitive, and the competition derives from both domestic and international sources. Twenty-three percent of product shipments go to exports, and imports account for 21 percent of apparent consumption. Given this type of market profile, survival demands that U.S. producers continue to operate at the leading edge of technology.

Industrial Restructuring

Several trends seem to point toward an eventual restructuring of the industry. As technology in the industry has advanced, it has become more complex and thus more costly to undertake the design and manu-

Table 5. U.S. Captive Suppliers, 1980

Company	Location
Aerojet Electro Systems	Azusa, California
Amdahl Corporation	Sunnyvale, California
Ampex Corporation	Santa Monica, California
Bell Telephone Labs	Murray Hill, New Jersey
	Allentown, Pennsylvania
Boeing Company	Seattle, Washington
Burroughs	San Diego, California
Chrysler Corporation	Huntsville, Alabama
Control Data Corporation	Bloomington, Minnesota
Cutler-Hammereaton	Milwaukee, Wisconsin
Data General	Sunnyvale, California
Datel Systems	Canton, Massachusetts
Delco Electronics Division	Kokomo, Indiana
Digital Equipment Corporation	Worcester, Massachusetts
	Hudson, Massachusetts
Eastman Kodak	Rochester, New York
E Systems, Inc.	St. Petersburg, Florida
John Fluke Mfg.	Everette, Washington
Ford Aerospace Communications	Newport Beach, California
Foxboro Company	Foxboro, Massachusetts
	San Jose, California
Four-Phase System, Inc.	Cupertino, California
General Dynamics	Fort Worth, Texas
General Electric	
SSAO	Syracuse, New York
Corporate R&D	Schenectady, New York
Aerospace Electronics Systems	Utica, New York
Gould, Inc.	Rolling Meadows, Illinois
GTE Laboratories	Waltham, Massachusetts
Hewlett-Packard	
Cupertino IC Operations	Cupertino, California
Santa Clara Division	Santa Clara, California
Microwave	Santa Rosa, California
Microwave Semiconductor Division	San Jose, California
H. P. Laboratories	Palo Alto, California
Optoelectronics	Palo Alto, California
Instrument Division	Colorado Springs, Colorado
Desktop Computer Division	Fort Collins, Colorado
Instrument Division	Loveland, Colorado
Handheld Calculator Division	Corvalis, Oregon
Honeywell	
Solid State Electronics Center	Plymouth, Minnesota
Solid State Electronics Center	Colorado Springs, Colorado

Source: William I. Strauss, ed., Status '81 (Scottsdale, Ariz.: Integrated Circuit Engineering, Inc., 1981),

R&D Lab	Proto-type Lab	Pilot Pro-duction	Full Pro-duction	Comments
x				
x	x			E-beam capability
			x	Bipolar
x	x			Production by Western Electric
	x	x		Co-located with Western Electric
x	x			Bipolar
x	x	x	x	Expanding
x	x			Planned
x	x	x	x	Limited open market supply
x				Bipolar R&D
x	x	x	x	MOS and bipolar std. production
	x			
x	x	x	x	Very large production for GM
x	x	x	x	Being phased out
x	x	x	x	Bipolar and MOS
x	x	x	x	Specialized circuits
x	x			
x	x			Operational January 1981
x				Imaging
x				Silicon sensors
x	x			Planned
x	x	x	x	Custom and std. MOS products
		x		CMOS
x	x	x		Prototype support
x	x			Basic R&D
x	x	x	x	Military production
x	x			
x	x	x		
x	x	x	x	CMOS/SOS
x	x	x	x	Bipolar
x	x	x	x	GaAs
x	x	x	x	Specialty devices
x	x	x		Bipolar, MOS R&D
x	x	x	x	LED and bipolar
x	x	x	x	Bipolar
x	x	x	x	NMOS
x	x	x	x	JFETs, bipolar, thin film
x	x	x	x	CMOS and NMOS
x	x	x		
			x	Production facility

table 5, continued

Company	Location
IBM	
Corporate	Yorktown Heights, New York
General Systems Division	Rochester, Minnesota
General Technology Division	Hopewell Junction, New York
General Technology Division	Essex Junction, Vermont
System Development Division	San Jose, California
Data Products Division	Tucson, Arizona
General Systems Division	Austin, Texas
Federal Systems Division	Manassas, Virginia
	Sindelfingen, West Germany
	Corbeil Essonnes, France
	Boeblingen, West Germany
	Ruschlikon, Switzerland
	Yasu, Japan
Lockheed Missiles and Space	Sunnyvale, California
Magnavox	Fort Wayne, Indiana
Martin Marietta Aerospace	Orlando, Florida
McDonnell Douglas Astronautics	Monrovia, California
(formerly Actron Labs)	St. Louis, Missouri
Micro Rel	Tempe, Arizona
NCR	Miamisburg, Ohio
	Colorado Springs, Colorado
	Fort Collins, Colorado
Northern Telecom	San Diego, California
	Ottawa, Canada
Northrop	Palos Verde, California
Rosemount	Eden Prairie, Minnesota
Sandia Laboratories	Albuquerque, New Mexico
Sperry	
Sperry Univac	St. Paul, Minnesota
Sperry Research	Sudbury, Massachusetts
Storage Technology Corporation	Louisville, Colorado
STC Microtechnology Corporation	Sunnyvale, California
Stomberg Carlson	Rochester, New York
Tektronix	Beaverton, Oregon
UTI Essex Group, Inc.	Pittsburgh, Pennsylvania
Western Electric	Allentown, Pennsylvania
	Reading, Pennsylvania
Teletype Corporation	Skokie, Illinois
Westinghouse	
Friendship Solid State Research	Baltimore, Maryland
	Pittsburgh, Pennsylvania
Xerox	El Segundo, California
Parc	Palo Alto, California

R&D Lab	Proto-type Lab	Pilot Pro-duction	Full Pro-duction	Comments
x				R&D
	x			Prototyping
			x	East Fishkill Production Facility
	x	x	x	Burlington Production Facility
x	x			
		x	x	Production facility
		x	x	Production facility planned
x	x	x		Military-oriented
x	x	x		
		x	x	
		x	x	
x	x	x		
x				
x	x	x		
x		x		
x	x			
x				
x				
x	x			
x	x	x	x	MNOS and logic
x	x	x	x	MNOS, RAM, custom logic
			x	CMOS, custom logic
x	x	x	x	On line 1981
x	x	x	x	
x				
x				
x	x	x		Non-profit laboratory
x	x	x		
x				
x				
x	x	x		
x	x	x		
x	x	x	x	
			x	PMOS products
	x	x	x	Microprocessors, RAMs
			x	Bubble memories
	x	x	x	
x	x	x	x	
x				
x	x	x		
x	x			Expanding

Table 6. U.S.-Based Integrated Circuit Manufacturers, World Sales of Firms with Sales More Than $10 Million (in $ millions)

Supplier	1979		
	MOS	Bipolar	Total
AMD	$95	$112	$207
AMI	94	--	94
Analog Devices	26	39	65
Cherry Semiconductor	--	8	8
Commodore	28	--	28
Electronic Arrays (NEC)	14	--	14
Exar	--	15	15
Fairchild (Schlumberger)	99	266	365
General Instrument	90	--	90
GTE Microcircuits	20	--	20
Harris	45	75	120
Hughes	14	2	16
Intel	430	30	460
Intersil	57	24	81
Micro Power Systems	9	3	12
Monolithic Memories	--	37	37
Mostek (UTI)	220	--	220
Motorola	290	220	510
National	210	280	490
Precision Monolithics	--	23	23
RCA	85	80	165
Raytheon	--	32	32
Rockwell	105	--	105
Solid State Scientific	26	--	26
Signetics	50	215	265
Silicon General	--	12	12
Siliconix	28	8	36
Sprague	--	23	23
STD Micro Systems	15	--	15
Supertex (Honeywell)	7	--	7
Synertek	50	--	50
Teledyne	9	6	15
Toshiba (Maruman)	12	--	12
TRW	--	18	18
Texas Instruments	350	557	907
Western Digital	14	--	14
Zilog	18	--	18
Others	20	15	35
TOTALS	$2,530	$2,100	$4,630

Source: William I. Strauss, ed., Status '81 (Scottsdale, Ariz.: Integrated Circuit Engineering, Inc., 1981),

1980/ 1979 Growth	1980		
	MOS	Bipolar	Total
39%	$135	$152	$287
19	112	--	112
15	32	43	75
25	--	10	10
54	43	--	43
100	28	--	28
20	--	18	18
23	113	337	450
33	120	--	120
25	25	--	25
54	50	135	165
50	22	2	24
34	570	45	615
19	67	29	96
25	12	3	15
138	--	88	88
66	365	--	365
34	395	290	685
47	295	425	720
39	--	32	32
21	95	105	200
38	--	44	44
10	115	--	115
27	33	--	33
47	55	335	390
50	--	18	18
22	35	9	44
35	--	31	31
20	18	--	18
43	10	--	10
20	60	--	60
20	12	6	18
25	15	--	15
72	--	31	31
36	460	770	1,230
93	27	--	27
50	27	--	27
60	31	25	56
37%	$3,377	$2,983	$6,360

Table 7. Market Shares in U.S.-Japanese-European Semiconductor Market ($ billions)

	1979	World Market Share	1980	World Market Share
Semiconductors				
United States	$5.037	43.7%	$6.326	44.8%
Japan	3.379	29.3	4.192	29.7
Europe	3.089	26.8	3.576	25.3
Total	11.505		14.095	
Integrated Circuits				
United States	$3.636	48.6%	$4.804	49.8%
Japan	2.072	27.7	2.680	27.8
Europe	1.764	23.7	2.152	23.4
Total	7.462		9.636	

Source: "World Market Forecast," Electronics, 13 January 1981.

facture of integrated circuits. The pace of entry into the industry slowed substantially during the 1970s due to the increased cost of start-ups and a contraction in the venture capital market. The venture capital market has more recently expanded due to the reduction in the maximum capital gains tax rate to 28 percent under the Tax Reform Act of 1978.[26] In 1979, $900 million of venture capital flooded the market, up from only $10 million in 1975.[27] The majority of this capital appears, however, to be going to existing firms that are expanding production under established technologies; little capital is being used to fund those firms, mostly new, that are exploring innovative technologies.[28]

Despite the revitalization of the venture capital market, there has been a strong trend toward acquisition. Of thirty-six new firms created from 1966 to 1979, only seven remained independent corporations in 1981.[29] Many factors motivate such acquisitions. The owners of some small firms actually establish their companies in order to realize the capital gains often associated with such takeovers. In other cases semiconductor firms unable to raise sufficient capital through the market

Table 8. Acquisitions of Independent Semiconductor Makers, 1975–1979

Year	Acquisition	Buyer	Price ($ millions)
1975	Signetics	U.S. Philips Trust, Netherlands	49
1976	MOS Technology	Commodore International, U.S.	1
1977	Litronix	Siemens, West Germany	16
	20%, Advanced Micro Devices	Siemens, West Germany	27
	Solid State Scientific	VDO, West Germany	5
	24%, Intersil	Northern Telecom, Canada	11
1978	Electronic Arrays	Nippon Electric, Japan	9
	Spectronics	Honeywell, U.S.	3
	Synertek	Honeywell, U.S.	24
	25%, American Microsystems, Inc.	Bosch, West Germany	14
1979	Mostek	United Technologies, U.S.	349
	Microwave Semiconductor	Siemens, West Germany	25
	Fairchild Camera	Schlumberger, France	397
	14%, Unitrode	Schlumberger, France	10

Source: "Can Semiconductors Survive Big Business?" Business Week, 3 December 1979, 67.

or internal earning have allowed themselves to be acquired by larger corporations in order to finance expansion. On the other hand, acquired firms may run into the problems of technological performance, including an apparent inability to innovate.[30]

Many acquiring firms have integrated backwards to gain a steady supply of semiconductor devices (table 8 details acquisitions from 1975 through 1979). Some firms have in turn undertaken forward vertical integration, stressing product development and market expansion over technological innovation. As semiconductor manufacturers are finding more numerous applications for their products in consumer and industrial goods, they have penetrated the mass market. In 1976 only one company was using more than $100 million of semiconductors annually; in 1979 there were seven such firms, and the total was expected to reach seventeen in 1981.[31]

These trends toward mass production and increased cost of design and manufacture lead many analysts to conclude that new firms will find it very difficult to enter the industry. Furthermore, it seems likely that established firms will increase their market concentration and may therefore be able to influence the pace of technological innovation.

Industry Employment

The structure of employment in the microelectronics industry differs among its three phases of product development and manufacture. Product development is synonymous with the research and development or design phase of production. It involves "(1) original conception (which includes ascertaining that a market exists), (2) extremely complex design and engineering of circuits to meet given product specifications, and (3) prototype production in order to refine and correct the design and to adjust the manufacturing process so that it will produce satisfactory yields. Only then is the product ready for production."[32] The second phase of production is known either as wafer fabrication or advanced manufacturing. In this phase chemical and physical processes are used to embed a circuit design in silicon wafer from 2 to 10 centimeters (1 to 5 inches) in diameter. Hundreds of individual circuits a half-centimeter (quarter-inch) square can be formed on each silicon wafer. The circuits are tested by machine, and then each circuit is mechanically cut out from the wafer. Circuits that pass tests for quality control then progress to the third and final phase of production, known as assembly or bonding. Each circuit is encased in a plastic or ceramic package, and tiny wires are welded to the circuit and to metal alloy leads or pins embedded in the package. These packages can then be plugged into circuit boards to perform their required functions.

As a whole, the industry is distinguished by its dependence on a highly skilled scientific and engineering work force. Forty percent of the total domestic semiconductor work force in 1972 were nonproduction workers, compared to 16 percent in the automotive industry and 13 percent in the apparel industry. The proportion of skilled to unskilled workers decreases as production changes from the design phase to the assembly phase. The ratio of skilled to unskilled workers in wafer fabrication is approximately 1:2. The assembly phase requires almost exclusively unskilled workers.

The assembly process is also highly repetitive, requiring little technology, equipment, or skill. Because it costs little to transport the lightweight circuits, even by air, semiconductor firms usually conduct the great proportion of final assembly in areas where labor is extremely cheap. Most employ about one-half of their total labor force outside the United States, chiefly in Asian countries. American corporations based in California have paid wages as low as $30 per month in Indonesia and $90 per month even in Taiwan.[33]

Cheap labor is by far the most prominent incentive for offshore assembly, but other incentives do exist. Many countries offer tax exemptions, liberal foreign exchange rules, and allowance of 100 percent foreign ownership to foreign employers, and several countries have provided modern infrastructure for foreign firms in export processing zones where production is also tariff-free. Currently, Scotland and Ireland are in stiff competition to attract U.S. and Japanese semiconductor assembly and wafer fabrication facilities, for products aimed at the Common Market. By producing in Common Market countries, U.S. and Japanese firms would escape the Common Market's 17 percent tariff on semiconductor imports; in addition they have been offered a variety of tax incentives, cash grants, skilled labor provision, and academic connections.[34]

The U.S. Tariff Schedule also creates an incentive for semiconductor manufacture overseas. Items 806.30 and 807 specify that manufacturers must pay import duties only on value added abroad, excluding the original export value and overhead included in the final price. In 1978 the semiconductor industry reimported more duty-free goods than any other U.S. industry: $1,479 million, of which $887 million (60 percent) was duty-free.[35]

Because one of the major means of cutting production costs has been to cut labor costs, and because companies can conduct final assembly overseas at a fraction of the labor costs for comparable work in the United States, a large proportion of the assembly process will probably continue in developing nations. There is, however, reason to believe that past trends are not a particularly good guide to future developments within this industry. In the period from 1970 through 1973, the average value added through Asian assembly rose as a proportion of the value of the reimported product, from 44 percent to 55 percent, suggesting that a decreasing share of the cost of production was incurred within the United States. The change has been attributed to rising Asian wages and increased use of foreign subcomponents in offshore production. This same measure, however, decreased from 1976 through 1978, from 54 percent to 40 percent. Some analysts believe that this shift was due to the complexity and expense of producing sophisticated circuits, which has shifted more wafer fabrication and assembly back to the United States.[36]

Trends in Technological Change

Technological advances in the industry may change the location of production. The shift from large-scale integration (LSI) to very large-scale integration (VLSI) may move a substantial proportion of circuit assembly and packaging operations back to the United States from overseas. VLSI circuits are more difficult to design than LSI circuits, and the testing of VLSI circuits is also more complex, requiring sophisticated equipment and powerful computers. The input-output connectors from a VLSI circuit to the pins in its package are more numerous than in a LSI circuit, requiring the use of sophisticated techniques of assembly. In addition, the packaging of VLSI circuits is larger and more expensive, which increases per unit shipping costs of assembled circuits.[37] Currently about 29 percent of assembly by U.S.-based firms is done in the United States, and market analysts predict that the proportion will rise to 39 percent by 1989. Reasons in favor of domestic assembly include (1) preference for conducting experimental operations near company headquarters, (2) protecting production from the threat of international transportation strikes, war, and civil unrest, (3) cost-effective domestic assembly of some circuits by more capital-intensive methods, and (4) federal prohibition of offshore assembly of military components.[38]

In line with the greater cost and complexity of VLSI, the production process is becoming less labor-intensive and more automated. Texas Instruments and IBM have both demonstrated a fully automated wafer

Table 9. Employment in the Semiconductor Industry, 1970-1977

Year	Total Employment	Production Workers	Production Workers/ Total Employment
1970	88,500	60,300	68.1%
1971	74,700	45,500	60.9
1972	97,600	58,400	59.8
1973	120,000	74,700	62.3
1974	133,100	81,600	61.3
1975	96,700	52,400	54.2
1976	102,500	59,700	58.2
1977	112,900	62,400	55.3

Source: U.S. Department of Commerce, Bureau of the Census, Annual Survey of Manufactures for 1970-1979 (Washington, 1978-83); Census of Manufactures, 1972, 1977 (Washington, 1976, 1981).

fabrication process.[39] The proportion of production workers in the industry (table 9) and the cost of payroll as a percentage of value of industry shipments (table 10) both decreased in the 1970s, indicating a reduction in the use of direct labor in semiconductor production. These statistics suggest that if much of the offshore assembly of semiconductors was due to the labor-intensive nature of their manufacture, then more recent trends in technology and automation will make domestic manufacture even more competitive with offshore assembly.

Another possible change in the organization of production is the "silicon foundry." A silicon foundry is a custom integrated circuit firm that manufactures circuits designed elsewhere. This arrangement may be optimal for end-users who can justify the addition to their operations of design capability but not manufacturing capability. This trend takes the division of the production phase one step further, to separate firms' handling of different phases of production.[40] The trend is also evident in offshore assembly, where the greatest growth recently has been in independent Asian assembly subcontractors.[41]

Table 10. Labor Intensity of Semiconductor Industry, 1970-1977

Year	Value of Shipments ($ millions)	Cost of Payroll ($ millions)	Cost of Payroll/ Value of Shipments
1970	1,501.2	663.6	44.2%
1971	1,599.6	658.1	41.1
1972	2,704.8	953.1	34.5
1973	3,647.7	1,201.0	32.9
1974	4,305.1	1,466.6	34.1
1975	3,276.9	1,200.5	36.6
1976	4,473.8	1,371.2	30.6
1977	5,238.2	1,587.5	30.3

Source: U.S. Department of Commerce, Bureau of the Census, Annual Survey of Manufactures for 1970-1979 (Washington, 1973-83); Census of Manufactures, 1972, 1977 (Washington, 1976, 1981).

Prospects and Problems

The nature of competition in the microelectronics industry is such that producers have to keep innovating just in order to stay even. Despite its current strength, the U.S. semiconductor industry will face problems maintaining its world market share and its lead in innovative capability against escalating foreign competition. The factors most often cited by the industry as impediments to future growth include (1) lack of adequate capital, (2) diminished support from federal research and development funding, (3) lack of skilled personnel in critical professions, and (4) regulatory bottlenecks impeding expansion to new markets.

Capital shortage. The microelectronics industry's earnings are not sufficient to finance its own phenomenal growth internally. From 1974 to 1979 there was a 31 percent decline in average return on equity and an 18 percent decline in the pretax profit margin; the industry's profit margin dipped again in 1980.[42] The increasing costs of new equipment are particularly burdensome for an industry with such high annual capital and research and development expenditures. Furthermore, the cost of capital in the United States has remained high relative to that in Japan, the leading competitor in microelectronics.

Table 11. Distribution of U.S. Semiconductor Sales by End Use

End Use	1960	1968	1974	1979
Computer	30.0%	35.0%	28.6%	30.0%
Consumer	5.0	10.0	23.8	27.5
Military	50.0	35.0	14.3	10.0
Industrial	15.0	20.0	33.3	37.5
Total value ($ millions)	$560	$1,211	$5,400	$10,500

Source: Robert W. Wilson, Peter K. Ashton, and Thomas P. Egan, Innovation, Competition, and Government Policy in the Semiconductor Industry (Lexington, Mass.: D. C. Heath, 1980), 19.

The estimated interest rate differential between Japan and the United States has ranged between 7 and 10 percent in recent years. These lower interest rates, in combination with the high debt to equity ratios permitted by the Japanese financial system, have given Japanese microelectronics firms a substantial cost-of-capital advantage.[43]

Federal support. To counteract the large-scale government subsidy of research and development in Japan and Europe, the U.S. industry is calling for a vigorous new effort in federal support. In fact, the Department of Defense has greatly expanded its financial support of research and development in the field of very high-speed integrated circuits (VHSICs); appropriations were $30.4 million for fiscal year 1980, with a six-year commitment to a $210 million program. Defense spending will still represent only about 10 percent of the industry's output, but it will increase approximately 23 percent per annum (see table 12).[44] The importance of the military market to the industry has declined, while the consumer and industrial markets have gained considerably.

Many analysts believe that the U.S. government will eventually have to provide direct funding for long-term research with commercial objectives in order for domestic firms to compete successfully with foreign microelectronics firms. The reasons for this are twofold: first, the governments of France, Germany, the United Kingdom, and Japan have all made major public investments in commercially oriented electronics research programs; second, the requirements of military electronics technology have become increasingly divergent from commercial applications. Hence one of the chief bases for U.S. predominance in microelectronics—the long-term, rising demand for semiconductor devices by the Department of Defense and the National Aeronautics and Space Administration—has begun to erode significantly.[45]

Shortage of skilled labor. In a 1980 survey of the microelectronics industry, industry officials labeled the shortage of skilled workers as the bottleneck mostly likely to impede future growth.[46] To help customers design microprocessor applications into products, the industry is forced to supplement its microelectronics engineering skills, on which it has based its growth, with software engineering skills. There thus has arisen the dual problem of a shortage of the technicians and engineers directly related to production as well as a shortage of those who design applications for the hardware. Both types of shortages could reduce the pace of growth in the industry.

Between 1972 and 1977 the annual production of electrical and electronic engineering graduates in the United States actually fell, from 17,632 to 14,085. More recently, the number of bachelor's degrees awarded in engineering has increased, but the growth in supply has not met current demand. Doctorates have continued to decline, which has adversely affected university engineering programs and, more generally, the progress of basic research.[47] The shortage of computer scientists is apparently even more serious. Rapidly rising undergraduate enrollments have not come close to meeting current demand. There were estimated to be 54,000 job opportunities available to graduates with bachelor's degrees in computer science in 1981, but only 13,000 graduates that year. As with engineering students, few undergraduate computer scientists go on to seek graduate degrees, leaving university computer science departments in a critical predicament.[48]

In many states there are major shortages in skilled engineering and science technicians, the workers who provide crucial support for engineers and scientists in the microelectronics industry. According to many observers, the institutions that train these individuals all too frequently have outdated, if not obsolete, equipment and poorly trained faculty in the necessary specialities. In most states the training of these technicians is a wholly haphazard enterprise, involving various combinations of technical institute training programs, vocational training in technical schools, on-the-job training, and educational programs in the armed forces. Few states with large microelectronics industry sectors, or the prospect of major growth in these sectors, have allocated adequate financial resources to programs to train engineering and science technicians.

Regulatory hurdles. In general, the microelectronics industry is not extensively regulated. The few existing problems of regulatory impediments can be viewed as a subset of the larger question of access to new markets. Microelectronics producers seek constantly increasing rates of incorporation of components into products and processes. The 1980 survey referred to above outlines two examples, one in banking and one in telecommunications.[49] The McFadden Act of 1972 requires that banks abide by the laws of the state of location and furthermore forbids banks to operate across state lines. Only a national venture, with a rational system of regulations, can open up the possibility of a cash-less/checkless society, making electronic funds transfer a reality. In the area of telecommunications, there is a huge but still largely untapped market for microelectronic devices. The computer and telecommunications industries are increasingly converging. AT&T, for example, has proposed plans for an "Intelligence Network," named Advanced Communication Service (ACS), which would allow terminals made by different manufacturers to communicate with each other; ACS would make the necessary signal translations. However, a condition of allowing AT&T to dominate telecommunications has been a *bar* on the corporation's competing in other industries. The question turns on where one draws the line between what is a piece of telephone equipment and what is a piece of computing equipment. This blurring of old distinctions between supplier and final-goods industries, or across different types of industry functions, will become more prevalent with increasing incorporation of microelectronics technology into products and processes.

Business climate. Representatives of the microelectronics industry, in accord with most business interests, have felt that government policy in recent years has been largely responsible for decreased capital availability, declining innovation, and other aspects of poor business performance. The Semiconductor Industry Association (SIA) supported the federal package of tax law revisions and depreciation schedule changes. It supported the new 10–5–3 depreciation schedule but would have preferred a three-year depreciation schedule for machinery. It also supported the reduction in the maximum capital gains tax from 28 percent to 21 percent and the special 25 percent tax credit on research and development expenditures that are above the average for such expenses over the prior three years. Industry officials believe that these revisions will stimulate investment—that, in effect, they more accurately reflect the reality of operations in an advanced-technology, rapidly changing industry.

Notes

1. In 1978 production in color television accounted for more than 52 percent of the value of shipments of all industries in the subsector. Unit output of color televisions rose by 8.6 percent from 1978 to 1979, and by 17.0 percent from 1979 to 1980. Over the same interval there was a sharp decline in the other segments of consumer electronics: black-and-white television sets registered yearly declines in unit output of 2.9 percent, then 7.5 percent; stereo and hi-fi equipment declined by 5.0 percent, then 10.0 percent. See U.S. Department of Commerce, *U.S. Industrial Outlook, 1980* (Washington, D.C., 1980), 397–403, and *U.S. Industrial Outlook, 1981* (Washington, D.C., 1981), 447–63.

2. See Governor's Task Force on Science and Technology, report, vol. 2: *Economic Revitalization through Technological Innovation* (Raleigh: North Carolina Board of Science and Technology, 1984).

3. Cited in *USA Today*, 21 January 1984, 1.

4. See, for example, Colin Norman, *Microelectronics at Work: Productivity and Jobs in the World Economy*, Worldwatch Paper 39 (Washington, D.C.: Worldwatch Institute, October 1980).

5. See "Commerce Department Sees a 7% U.S. Electronics Growth, to $97.6 Billion Industry in 1980," *Electronic News*, 11 February 1980; see also Norman, *Microelectronics at Work*, 40–51.

6. Arthur L. Robinson, "Perilous Times for U.S. Circuit Makers," *Science*, 9 May 1980.

7. Jerry Saunders, president of Advanced Micro Devices, quoted in Norman, *Microelectronics at Work*, 13.

8. See, for example, William I. Strauss, ed., *Status '81* (Scottsdale, Ariz.: Integrated Circuit Engineering, Inc., 1981), 36: "The 1970s introduced the computer revolution, with main frame horsepower becoming available in minicomputers and even desktop computers. Advances in integrated circuits were the driving force behind this revolution. This decade will make the last one seem almost insignificant in comparison. The horsepower that is now only available in big machines like Cray and Cyber will be at the fingertips of almost every engineer. Tremendous computing power will even be available in the household by the end of the decade."

9. Ernest Braun and Stuart MacDonald, *Revolution in Miniature: The History and Impact of Semiconductor Electronics* (Cambridge: Cambridge University Press, 1978), esp. chap. 4.

10. Braun and MacDonald, *Revolution in Miniature*, chap. 10.

11. See, for example, Annalee Saxenian, "Silicon Chips and Spatial Structure: The Industrial Basis of Urbanization in Santa Clara County, California" (Working Paper 345, Institute of Urban and Regional Development, University of California, Berkeley, March 1981).

12. Braun and MacDonald, *Revolution in Miniature*, chap. 10.

13. Saxenian, "Silicon Chips and Spatial Structure."

14. Saxenian, "Silicon Chips and Spatial Structure."

15. Strauss, *Status '81.*

16. See Department of Commerce, *U.S. Industrial Outlook, 1981,* 305–15.

17. John Young, president of Hewlett-Packard, quoted in Strauss, *Status '81.*

18. See Nico Hazewindus, *The U.S. Microelectronics Industry: Technical Change, Industry Growth and Social Impact* (New York: Pergamon Press, 1982), 51.

19. See Strauss, *Status '81.* Major captive firms include divisions and subsidiaries of IBM, DEC, Amdahl, Burroughs, Honeywell, Control Data, Sperry, and General Electric.

20. Strauss, *Status '81.*

21. See, for example, Braun and MacDonald, *Revolution in Miniature,* chap. 11.

22. Strauss, *Status '81.*

23. See Robert W. Wilson, Peter K. Ashton, and Thomas P. Egan, *Innovation, Competition, and Government Policy in the Semiconductor Industry* (Lexington, Mass.: D.C. Heath, 1980), 158–64.

24. See National Academy of Engineering, *The Competitive Status of the U.S. Electronics Industry* (Washington, D.C.: National Academy Press, 1984), 28–38, 114–16.

25. Braun and MacDonald, *Revolution in Miniature,* chap. 11.

26. "Report Increase in Venture Capital," *Electronic News,* 20 July 1981.

27. "Singing the Semiconductor Blues," *New York Times,* 24 May 1981, F4.

28. See Wilson, Ashton, and Egan, *Innovation, Competition, and Government Policy,* 164–70.

29. Saxenian, "Silicon Chips and Spatial Structure."

30. Saxenian, "Silicon Chips and Spatial Structure."

31. Saxenian, "Silicon Chips and Spatial Structure."

32. Saxenian, "Silicon Chips and Spatial Structure."

33. Lenny Siegel, "Delicate Bonds: The Global Semiconductor Industry," *Pacific Research* 9, no. 1 (1980).

34. Martin Gold, "Scotland, Ireland Vie for World Semicon Maker; Wafer Pre-empting Wool, Whiskey as Battle Escalates," *Electronic News,* 3 August 1981.

35. Siegel, "Delicate Bonds."

36. Siegel, "Delicate Bonds."

37. Strauss, *Status '81.*

38. Siegel, "Delicate Bonds."

39. Strauss, *Status '81.*

40. Strauss, *Status '81.* See also F. Dana Robinson, chapter 3, this volume.

41. Siegel, "Delicate Bonds."

42. "Semiconductors: Fighting Off the Japanese," *The Economist,* 7 June 1980; Department of Commerce, *U.S. Industrial Outlook, 1981,* 447.

43. National Academy of Engineering, *Competitive Status,* 3–5, 108–10.

44. Strauss, *Status '81.*

45. National Academy of Engineering, *Competitive Status,* 105–8.

46. "Microelectronics: A Survey," *The Economist*, 1 March 1980 (a separately paginated supplement). In the same year Intel's chief officer Gordon Moore said he was far less concerned about the roller-coaster economy than he was with getting enough skilled workers to support the industry's future growth; see "Rolling with the Recession in Semiconductors," *Business Week*, 21 July 1980, 189, 192.

47. National Academy of Engineering, *Competitive Status*, 17–20.

48. National Academy of Engineering, *Competitive Status*, 17–18.

49. "Microelectronics: A Survey."

Innovation in Microelectronics: Implications for Economic Development Planning

F. Dana Robinson

"Microelectronics" covers as many as two hundred highly diverse firms in the United States that are involved in some aspect of the production of advanced integrated circuits (ICs), the tiny transistors and other electronic components embedded in silicon chips. Despite the disparities within an "industry" that includes IBM, General Motors, Intel, and numerous small firms, their common concern with the miniaturization of components does make it meaningful to identify common aspects of their use of microelectronics technology. Many highly miniaturized integrated circuits are standardized products, especially the computer memory chips that are the industry's best-selling product. Competition between the leading American and Japanese semiconductor producers in the memory market has received extensive press coverage, along with related issues like the capital formation problems facing some U.S. producers. This chapter examines a less publicized aspect of the production of highly miniaturized integrated circuits.

Miniaturization allows single chips to function as more or less self-contained electronic systems. Though memory chips have regular configurations and are dedicated to a single function, many innovative chips are extremely complex, multipurpose systems that can be custom-designed for particular applications in offices, factories, telecommunications systems, and military equipment. Because these chips are not standardized, their production is not in the hands of the dozen firms in the United States (mainly in Silicon Valley) and Japan that are responsible for almost all innovations in memory chips. Instead the number of firms involved in innovations is larger, as is their geographical dispersion.

The following analysis of the implications of this situation is presented in terms intended to be useful to local economic development planners. The task of developing innovative, special-purpose integrated circuits can be divided up among different segments of the microelectronics industry. The spatial implications of this division of labor are considered, along with possible effects of future technological advances on these patterns, and the implications for economic development planning.

The Organization of the U.S. Microelectronics Industry

The first semiconductor devices, which came onto the market in the early 1950s, were individual transistors and diodes. When electronics firms used semiconductor components for computation, amplification, and other applications, they wired them together on circuit boards. Inventing the actual circuits and forming the connections between components were the responsibility of user firms. The integrated circuits of the 1960s simplified the development of electronic circuits by combining up to several dozen transistors and other components onto a single chip. The results were fewer wires, much smaller circuits, and superior performance. The integrated circuits remained, however, building blocks to be plugged into larger circuits. As with transistors, integrated circuits were treated as components (generally purchased off the shelf) to be inserted into systems defined by users. The transistors, diodes, resistors, capacitors, and wires on today's integrated circuits are many times smaller than those of fifteen years ago, and hundreds of thousands can be fitted onto a single silicon chip. Today's most commercially important memory chips, 64K RAMs, can store more than 64,000 binary "bits" of information. Because the market for memory chips is so large, economies of scale have made 64K RAM prices comparable to those of previously available memories. Similar improvements in computing power and reductions in cost per function characterize chips for logic, process control, computation, and telecommunications applications.

Circuit designers, then, use advanced integrated circuits to lower the size of circuit boards, simplify wiring, and increase the speed and simplicity with which equipment can operate. One high-density chip costing perhaps $10 can now do all the work of eight less advanced chips costing a total of about $15 and requiring a printed circuit board that adds as much as $7.50 to the cost. A single $200 or $300 part for a computer now replaces up to 2,500 low-density chips worth $1 each. Moreover, operating speed using a single chip is greater, and equipment size and cooling requirements are reduced.[1]

Such improvements in performance and cost-effectiveness create new markets for semiconductor innovations. Peak sales of the 64K

RAM, for example, are forecast to exceed those of its predecessor, the 16K RAM, even though per-unit storage capacity is four times as great.[2] The implication is that there is an enormous demand for cheap computer power. Progress in miniaturization has also directly increased the ability of microelectronics innovations to create or further penetrate markets for logic, process control, computation, and telecommunications devices. Enhanced technological skill in packing components onto chips permits compact, low-cost, and high-speed electronic equipment to take over new functions (as in electronic control of engine performance) and to make computing power available to homes, small offices, and factories, similarly boosting microelectronics consumption.

However, in considering such applications, it is clear that miniaturization alone does not tell the whole story. If each chip can function as a more or less independent system, or is responsible for a substantial proportion of the work done by a large-scale electronic system such as a mainframe computer, integrated circuits can no longer be treated as off-the-shelf components to be wired into a user's system. Today individual chips already contain much of a system's interconnections. Users relying on the open market for integrated circuits necessarily find that important decisions concerning the precise functioning of the user's system will already have been made by the chip manufacturer. The manufacturer may not have designed the chip in exact accord with the user's own particular requirements. In this sense the microelectronics industry's organization is in part determined by decisions as to which firm controls the configuration of chips' circuitry. The chips sold on the open market reflect configurations that have been chosen by semiconductor manufacturers rather than users. "Standard" chips are no longer standard in the earlier sense of being elemental building blocks for designing circuits.[3]

Users who want chips tailor-made to their own particular designs can control the design process in three ways. First, they can contract with large merchant producers such as Intel to have chips manufactured to specifications. However, access to such firms is generally limited to users that purchase integrated circuits in heavy volume. A second strategy is to manufacture the device internally through a "captive" semiconductor subsidiary. Because the start-up costs for such an undertaking are considerable and the range of devices available from independent semiconductor producers is extensive, decisions to create captive semiconductor lines are not taken lightly. Third, firms that need relatively small numbers of integrated circuits can still obtain custom chips by contracting with one of the many small "custom houses" specializing in such services rather than in volume production. In fact a wide variety of specialized services is available to microelectronics users (see table 1). Because of the comprehensive nature of these services, it is possible to intervene at different stages in the chip production process to gain the degree of customization required. The microelectronics industry is currently structured to allow considerable flexibility in determining chip characteristics. Users can intervene at critical stages in the design process without taking on full responsibility for manufacturing or conducting a full range of research and development.

Of total worldwide integrated circuit production by American firms, totaling $8.9 billion in 1981, production by captive subsidiaries for internal consumption amounted to $2.9 billion, and customized chips produced by independent semiconductor firms accounted for about $600 million in sales. However, the proportion of industry output represented by customized chips from both captive and independent sources is growing; independent production of custom chips could represent half of 1990 integrated circuit sales. Improvements in computer-aided design (CAD) are expected to make customized circuitry much easier to develop, reducing the economic attractiveness of more general-purpose chips.[4] Among independent suppliers both the large merchant firms and the more specialized custom houses can be expected to respond creatively to this trend, though the large merchant suppliers may lose market shares relative to other sources of supply. At least a dozen new specialized firms have opened in the 1980s, and electronics firms have established new captive facilities and expanded existing capacity.[5]

The Location of the U.S. Microelectronics Industry

Table 2 lists independent and captive microelectronics facilities by region. Some classifications are arbitrary, because the distinction between captive and open-market suppliers has begun to blur as parent firms expand the manufacturing capacity of their subsidiaries to levels that permit open market sales.[6] Conversely, many of the firms listed as open-market producers are owned by American or foreign electronics firms that may absorb part of their output.

Table 2 substantiates California's reputation as the heartland of the American microelectronics industry. Most firms manufacturing for the open market are located there, along with dozens of specialized service firms. In fact, California produces 34 percent of the world's

Table 1. U.S. Semiconductor Industry Services

Manufacturing

- custom IC supply
- wafer fabrication from users' design
- semicustom IC fabrication (alterations to standard hardware)
- hybrid fabrication (components not entirely integrated within a single chip)

Services to Low-Volume Chip Manufacturers

- architectural/engineering
- facility consulting
- equipment maintenance
- layout
- leasing/financing
- materials/gases
- microanalysis
- package assembly
- photomasks
- process (yield) consulting
- test program writing
- testing/screening
- venture capital

Chip Design

- circuit design
- IC design
- CAD services
- CAD software
- digitizing

Marketing Services

- advertising
- customs expediting
- manufacturing (sales) representation
- market reports
- market research

Source: Mel H. Eklund and William I. Strauss, eds., Status '82: A Report on the Integrated Circuit Industry (Scottsdale, Ariz.: Integrated Circuit Engineering Corporation, 1982), 27.

Table 2. Microelectronics Firms Classified by State and Type

	CA	OR WA ID UT	AZ CO NM
Open market IC mfg facilities (1981 sales > $10 million)	24 firms	6 branch plants	2 firms 5 branch plants
Other open market firms	27	1 (ID)	6
Captive IC facilities (R&D and/or mfg)	17	4	11
IC testing	26	2	4
Microanalysis	13		2
CAD software	15		
Design/consulting	19	1	4
Custom/semicustom/hybrid IC suppliers	23	1	3

Source: Mel H. Eklund and William I. Strauss, eds., Status '82: A Report on the Integrated Circuit Industry (Scottsdale, Ariz.: Integrated Circuit Engineering Corporation, 1982), appendix.

integrated circuits, more than the rest of the United States combined.[7] The relationship between the number of chip manufacturers and service firms is reciprocal. Obviously the presence of numerous manufacturers creates a strong demand for nearby services. Equally important from the standpoint of economic development planning, service firms make California a nurturing "incubator" for new chip producers. A twenty-year study by the University of Santa Clara that has tracked the progress of more than four hundred manufacturing start-ups has shown that the average manufacturing company has a 75 percent chance of surviving its first two years in business. By contrast, Silicon Valley companies in the study stood a 95 percent chance of surviving their first six years, reflecting the presence of a rich network of universities, support companies, consultants, and venture capitalists.[8]

California's advantages have not prevented major merchant producers such as Texas Instruments, Motorola, and Mostek from prospering

TX MO LA	MN	Great Lakes States	South[a]	New England	NY PA NJ MD
2 firms 3 branch plants			1 firm 1 branch plant	3 firms 2 branch plants	3 firms 1 branch plant
1			1	2	4
3	5	6	4[b]	10	13
4	1	3	2	11	6
	1	3	1	2	4
1				1	1
2	2		2[c]	2	7
1	2	2	1	4	8

[a]Florida, unless otherwise specified

[b]Two in NC

[c]One in AL

despite locations far from Silicon Valley. Moreover, constraints imposed by tightening supplies of housing, water, and room for expansion have contributed to recent decisions by firms like Fairchild and Intel to locate new facilities out of state. The value of close physical proximity to the California microelectronics industry is greater to the second tier of chip producers, those with sales of less than $150 million and more than $10 million. Of twenty-seven such firms, nineteen are located within the state.[9]

The captive portion of the microelectronics industry is somewhat less concentrated. Though four of the ten leading captive producers maintain facilities in California, important groupings of computer manufacturers with captive facilities can be found in Minnesota, Massachusetts, and the Middle Atlantic region. Manufacturers of other electronic and electrical equipment such as NCR, RCA, and General Electric have not located captive facilities in California. RCA and General Electric do maintain facilities distant from their headquarters in the Northeast, but they chose sites in Florida and North Carolina, respectively.

Regional concentrations of the microelectronics industry such as those in Minnesota, Illinois, Florida, and the Northeast do include specialized service firms, though in rather skimpy numbers by California standards. As the demand for customization increases with further integrated circuit component density, continuing start-ups and growth in this part of the industry can be expected.

Despite some geographical dispersion in the microelectronics industry, the distribution of activity remains very uneven. Important industrial centers like Pittsburgh, Cleveland, and Atlanta lack a microelectronics presence. Microelectronics firms are sparse in the Southeast (excluding Florida) and, to a lesser extent, the industrial Middle West outside Chicago and Minneapolis. Historically, the location of microelectronics activity has reflected three different types of decision criteria depending upon the firm involved.

First, as the Santa Clara University study suggests, entrepreneurs need ease of access to outside technical and financial services and also a milieu in which close personal contacts can be maintained with customers, potential employees, and colleagues working in related technical areas. The initial establishment of Silicon Valley as a location offering these advantages (and other amenities as well) was due to William Shockley, an inventor of the transistor, who had grown up in Palo Alto. He decided to locate the Shockley Semiconductor Laboratories there in 1954 after leaving Bell Laboratories. Once an electronics industry was established in the Bay area, those of Shockley's employees who left to start up new firms tended to remain in the vicinity.[10] As it developed in the 1950s the transistor industry was dominated by large, established electrical manufacturers such as RCA, Sylvania, Westinghouse, and General Electric. California did not then account for the major portion of semiconductor production or research and development. By the late 1960s, when integrated circuits had become the most important product of the industry, this initial group of firms had been forced to the margin of the industry by newer, more entrepreneurial firms that had been entering the semiconductor industry in large numbers. Several withdrew entirely. This greatly reduced the share of activity taking place outside California, Texas, and Arizona.

Second, during the 1970s improvements in integrated circuit performance and cost-effectiveness spurred a tremendous increase in the range of microelectronics-based applications. The interdependent relationship between innovations in microelectronics and final equipment is now well recognized. This trend, accompanied by fierce competition between numerous electronics firms in the United States and Japan, has made expertise in microelectronics more and more essential to equipment manufacturers. Thus the number of captive facilities, service firms, and smaller independent firms (often custom houses) has increased in the areas outside California where equipment manufacturers are located. Manufacturers have generally not placed captive facilities in locations that do not already support electronics equipment manufacturing.

Third, the largest merchant and captive producers may not always find it desirable to keep all facets of chip development, production, and assembly into final equipment in close geographical proximity. For example, for relatively routine fabrication and assembly of chips, location in an area of low labor costs may have important economic advantages. The largest merchant producers maintain low-skill facilities abroad, and such considerations have also hastened the spread of integrated circuit production into Oregon, Washington, and northern New England. Selecting a site for a low-wage branch plant in microelectronics is not very different from locating branch manufacturing facilities generally.[11]

Acquisitions in the Microelectronics Industry

The past decade has seen the acquisition of about two dozen independent U.S. merchant integrated circuit producers, the largest being Fairchild, by electronics firms in the United States and abroad. This depletion of the ranks of the independent producers, which has caused some commentators to worry about the loss of entrepreneurial management styles in the semiconductor industry,[12] has actually not led to major shifts in their orientation to markets. Acquired firms have not, for the most part, become captive subsidiaries of the parent firms.

By purchasing chip producers, a firm acquires not only a stake in a fast-growing industry but also an inside look at technological developments and a source of in-house microelectronics expertise. Vertical integration—in the sense of functional links between "upstream" development and manufacture of chips and "downstream" development of equipment and consumption of chips—has not necessarily resulted from such acquisitions. Nevertheless, managers of research and development in the parent firm can draw upon the experience and research capabilities of their counterparts in the new subsidiary. In this sense many American electronics manufacturers are moving toward the kind of organizational capabilities enjoyed by both IBM and Japanese electronics giants like Hitachi and NEC.

Capital formation problems in financing semiconductor investments do not preoccupy the large electronics firms with independent or captive microelectronics subsidiaries as they do the major merchant producers. A successful firm with a commitment to staying in the merchant market over the long haul is in a position to secure the kind of long-term financing for its subsidiary that is often elusive to smaller independent producers. As for investments in internal integrated circuit supply capabilities, equipment producers find it somewhat easier to recover the capital costs of semiconductor facilities than do merchant producers, because competition in the markets they face does not drive prices down toward marginal costs with the same rapidity. The advantages of developing custom chips may also offset the cost of captive facilities. However, as with any firm seeking innovative applications of microelectronics technology, there are always risks that crippling technical snafus will occur, that personnel with key skills cannot be secured, and that the critical mass of activities making the facility cost-effective will prove elusive.[13]

In fact, the chief risk to the large firm that is integrating into upstream microelectronics capabilities may be difficulty in achieving the breadth and depth of expertise needed for rapid responses to technological change and developments in semiconductor and equipment markets. Such changes occur at a dizzying pace, and the ability to manage research and development programs under these conditions is not to be taken for granted. As a consequence, it has been the user firms with captive facilities that have initiated attempts to forge closer relations between university and industrial research and development. User firms have been the prime movers in research programs bringing personnel from industry to Stanford, California Institute of Technology, the University of Minnesota, and Rensselaer Polytechnic.[14]

Microelectronics Technology in the VLSI Era

The preceding discussion suggests that Japanese competition, rising investment requirements, and the quality of university programs have substantially different effects on different segments of the industry. The major merchant producers are bearing the brunt of overseas competition and tend to emphasize national issues such as trade policy, taxation, and the federal role in bolstering university research and training programs. The smaller semiconductor producers and the suppliers of specialized services do not compete as directly with Japanese firms and do not perceive as immediate a need for close links to universities. Their paramount concerns are likely to be the state of the venture capital industry and the impacts of the federal tax code on newly established firms. The locational decisions of such firms will hinge on easy access to customers and outside services. The captive portion of the industry is chiefly concerned with technical and management issues involved in exploiting the potential economies of scope that arise from interdependencies between innovations in microelectronics and innovations in finished electronic equipment. Because the typical equipment producer has long since discarded the management traits of entrepreneurial firms, an extra effort must be made to keep managers closely attuned to changes in the technological knowledge base so as to increase their capacity to respond to developments in microelectronics markets. Universities can supply some information on research activity taking place outside the firm; purchasing a semiconductor producer is another avenue for obtaining such knowledge.

Merchant firms may eventually have to concede a large portion of the memory chip market to the Japanese, but the range of technical possibilities for further innovation in the commodity chip markets is large. U.S. merchant firms may well create new products in which they maintain dominant market shares. At the same time, as the pervasive diffusion of microelectronics into new applications increases the demand for customized products, the relative importance of high-volume, general-purpose chips is likely to decline. In turn, the role of captive producers and their suppliers will become more prominent. The implications of some of these possibilities are explored below.

Since 1964, when commercial sales of integrated circuits first became sizable, successive innovations in the components making up integrated circuits, improvements in fabrication machinery, and superior processing techniques have made it possible for firms to shrink component size further and further, thereby allowing chips of a given size to contain more and more computing power, memory capacity, signal processing capability, and other electronic functions. This miniaturization has proceeded without interruption, and at a rather regular (and rapid) pace. In 1965 Gordon Moore, then the director of research at Fairchild and later a founder of Intel, predicted that the density of components on the most advanced integrated circuits would double every year. This actually proved to be the case over the next decade, though Moore later noted a "slowing"—to a rate of doubling in density every year and a half during the late 1970s.[15]

Technological progress in integrated circuits thus involves a predictable trajectory of regularly increasing chip density. The different levels of density along this trajectory are distinguished according to the number of components that can be "integrated" within a single integrated circuit. Technologies of the mid-1960s supported "small scale integration" of at most 100 transistors or other components per chip. The 64K RAM memory chip, with somewhat less that 100,000 transistors, represents a transition from "large scale integration" (LSI), going back to about 1970, to "very large scale integration" (VLSI), which will characterize the integrated circuits of the next few years.

The arrival of future generations of memory chips has been relatively predictable. Every three to four years, the capacity of the most advanced chips on the market quadruples. Strategies for competing in the memory market thus reduce to developing the means to manufacture chips whose components can be sufficiently shrunk to produce the required degree of density. State-of-the-art memories represent the maximum levels of circuit density permitted by the current degree of technological expertise in high-volume semiconductor fabrication. Consequently, successful innovation in design is heavily dependent on innovation in manufacturing processes.

Other VLSI chips involve the same manufacturing technology as memory chips but present additional areas of uncertainty about the direction being taken by innovation. A package of a few such integrated circuits can now be designed as a complete system for high-performance computers, real-time programmed control of industrial equipment, and advanced telecommunications and data transmission systems. But, in the words of one commentator, "The semiconductor industry regards VLSI design as its number one scientific challenge."[16] It is currently unclear how designers are going to keep abreast of the complexity of multifunctional integrated circuits in which the number of transistors doubles every few years.

Logic designers begin the process of connecting 100,000 or more individual components by creating a chip architecture that combines functional subsystems into a coherent system. Then a circuit designer converts the logic design into specific electronic circuits. A layout designer works out the placement of the individual components that permit circuits to function. Finally, the layout must be translated into masks (templates) which, after reduction to actual chip size, control integrated circuit manufacturing processes. With 100,000 components mistakes are inevitable, and designs must be modified during debugging. Even after prototypes have been fabricated, engineers may discover unexpected shortcomings in performance that necessitate further redesigning. For high-density chips, this process is time-consuming and expensive.[17]

In fact, VLSI design would be impossible without computer-aided design (CAD). CAD systems, available from more than a dozen specialized vendors, allow designers to use interactive computer graphics systems to define, visualize, and modify integrated circuit layouts. For defining actual layouts and converting them to the sizes needed for mask generation (about one fourth of the effort involved in the total design process), current CAD technology considerably reduces the labor involved. Verifying that the logic and circuit designs will function as needed, and debugging, which account for about two-thirds of the effort, have still not been extensively automated. CAD is thus as yet an incomplete aid to VLSI design problems; universities and private industry are racing to extend its scope. A number of different approaches for accomplishing this objective present themselves, but none has become established as the dominant trend in future CAD innovation.[18]

But devising a design strategy involves not only alternative CAD systems but also choices between fundamentally different types of integrated circuits. No single type of integrated circuit type is optimal for all applications, though innovative breakthroughs within a given approach may expand the range of applications relative to other types. This technological heterogeneity promises to reinforce the heterogeneity of organizations within the microelectronics industry.

The end user choosing integrated circuits for inclusion in an advanced piece of electronic equipment has a variety of options (see table 3). The ultimate choice reflects the nature of the customized microelectronics attributes that the user needs, (1) to create special performance features in its own product line, (2) to increase operating speed or lower equipment size, or (3) to lower total systems costs. These attributes may be created primarily through the development of software that has been customized to meet special user needs, through circuitry that embeds special functional capabilities in customized chips, or through specialized hardware that allows equipment to operate at desired speeds and under unusual electrical or environmental operating conditions.

When customized software suffices to create customized attributes, the user may be content with one of several types of memory chips that store software instructions. If more complex data processing is required in addition to the creation of a stored program, a relatively standard microprocessor with customized software may be cost-effective. When special hardware attributes, such as high speed, are sought, standard microprocessors may prove inadequate; redesigning the mi-

Table 3. IC Alternatives for End Users

1. Standard components:
   - Off-the-shelf medium- or large-scale integration ICs
   - User-programmable and -reprogrammable memories
   - User-programmable logic arrays
2. Microprocessors
3. Semicustom chips with final circuit configurations determined by user and added to semifinished standard chips:
   - Gate arrays
   - Transistor arrays
   - Analog component arrays
   - Read-only memories
4. Hybrid ICs
5. Fully customized ICs (requires CAD for VLSI)

Source: Mel H. Eklund and William I Strauss, eds., Status '82: A Report on the Integrated Circuit Industry (Scottsdale, Ariz.: Integrated Circuit Engineering Corporation, 1982), 33, 36.

croprocessor to fit particular hardware needs will be costly at high levels of complexity. Semicustom chips, involving regular groupings of components whose interconnections can be customized during the final stages of processing, are one alternative means for achieving customized hardware and software features without the expense of a fully customized chip. For applications that place unusual electrical demands on chips, hybrid devices may be developed in which some components are not integrated on the chip.

This broad range of options serves users that have varying software and hardware requirements. At present many users seeking customized chips in quantities that cannot justify heavy investments in circuit design tend to sacrifice hardware performance by relying on standardized programmable devices (microprocessors, memories, and logic arrays) or semicustom chips whose components are arranged in regular arrays rather than in configurations that maximize operating performance. However, as CAD capabilities improve, these users will increasingly find customized hardware cost-effective. Figure 1 suggests the market impact of progressive improvements in semicustom de-

Figure 1. Digital Integrated Circuit Market Trends

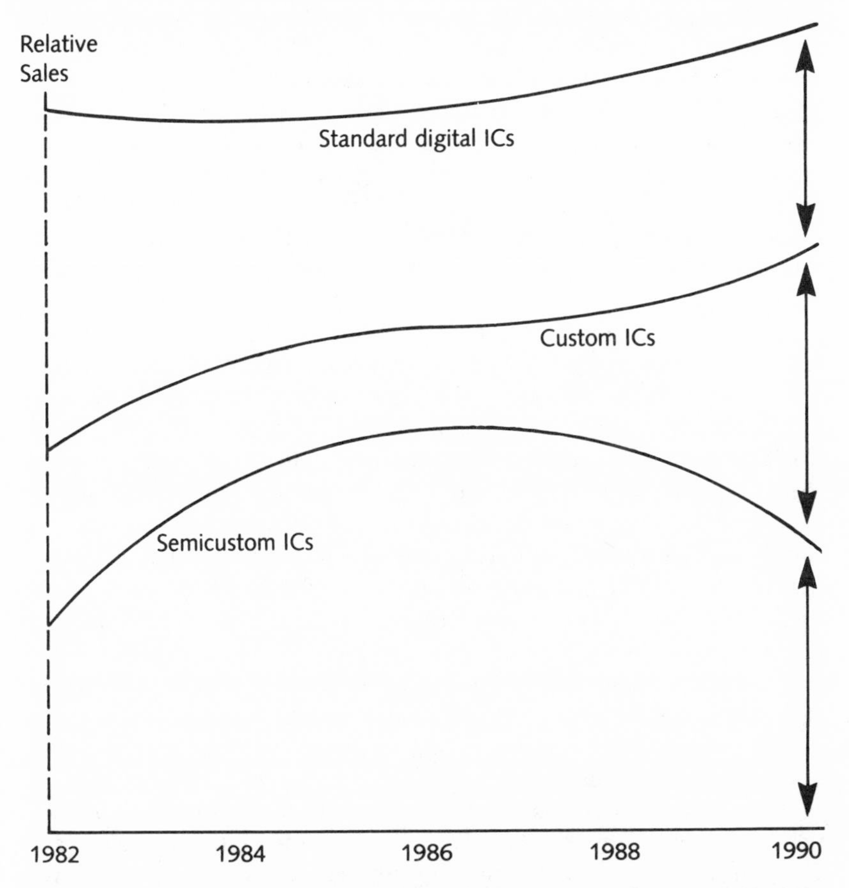

Source: Adapted from Mel H. Eklund and William I. Strauss, eds., *Status '82: A Report on the Integrated Circuit Industry* (Scottsdale, Ariz.: Integrated Circuit Engineering Corporation, 1982), 27.

vices, which are now coming into wide use, and in CAD, which is expected to permit largely automated design of custom integrated circuits by the end of the decade.

This glance at the wide range of possible design and hardware options for VLSI devices reinforces earlier observations that manufacturing commodity chips (such as memories) does not typify the microelectronics industry. To be sure, commodity chips will continue to dominate integrated circuit production. The semicustom market will demand increasing numbers of standardized, semifinished integrated circuits in whose development and manufacture the major merchant firms are in a strong position to lead. Improvements in programmable and reprogrammable memories will also benefit low-cost commodity producers. Despite such opportunities, major merchant firms will not necessarily dominate the development of the innovative design techniques that are becoming a prerequisite for VLSI innovations. Both the current popularity of semicustom integrated circuits and the anticipated impact of CAD toward the end of the decade signal a separation of chip design activities from manufacturing. The masks that incorporate customized features need not be developed by the manufacturer, but can be supplied by the user or a specialized service firm; CAD will soon allow user firms or their agents to generate, in cost-effective fashion, all the masks needed for manufacturing. The manufacturer can then serve primarily as a "silicon foundry" for manufacturing chips to specifications supplied by users.

When users approach major merchant producers with large orders, the latter may well aggressively seek to retain their current role in advanced chip design by offering customizing design services. However, the dynamic economies of scale that can be achieved in manufacturing, which give large producers major cost advantages in commodities markets, do not hold in design. Moreover, large capital investment is not necessarily required for the development of design expertise. VLSI thus presents opportunities to smaller firms that offer specialized design services and to captive facilities that wish to exploit the potential ability of customized VLSI chips to support greatly superior performance in user product lines.

University research programs offer a potentially important source of new ideas concerning CAD. For example, integrated approaches to CAD are being developed, in very different forms, at the California Institute of Technology and MIT. Private industry (particularly in the captive sector) has shown considerable interest in the progress of these research programs.[19]

Implications for State and Local Recruitment Policies

The multifaceted nature of the innovation process in microelectronics, and the diverse population of firms that has resulted, have significant policy implications for states and localities that wish to attract microelectronics firms. The successful, well-known semiconductor producers like Texas Instruments and Intel differ from small, open-market firms and from firms that produce mainly for internal consumption in finished electronics goods. Though activities of the first and second groups are heavily concentrated in California (Texas and Arizona account for much of the rest), the captive portion of the industry is much more widely dispersed around the country.

The technological determinants of success in the world markets in which the major merchant producers compete are not necessarily the same as those facing other firms in the microelectronics industry. In particular, the pursuit of technical solutions to problems in applications differs from the process of achieving greater density in memory chips. Moreover, different public policy issues are involved for the different segments of the industry. The major merchant firms appear to be primarily interested in the kind of support that the federal government can provide, such as changes in the taxation of high-technology firms and trade negotiations with Japan. Captive suppliers have a different public policy agenda.

This suggests that in states other than California, economic development planners should focus on the latter agenda. The captive portion of the microelectronics industry is not only more widely dispersed around the country, but it is also growing rapidly. It has likewise taken the lead in attempts to improve university research programs, and regional coalitions of such firms have focused on developing state-level initiatives.

Skilled manpower is *the* critical resource for all microelectronics firms, and this partially explains the efforts of captive producers to improve university training programs. Economic development planners should also keep in mind that captive producers inevitably encounter challenging issues on the frontier of the microelectronics knowledge base, particularly in chip design but also in manufacturing process techniques. Understandably, these firms seek every opportunity to monitor forthcoming technological advances. Nearby universities can provide such information if research is adequately funded.

Notes

1. "The Boom in Tailor-Made Chips," *Fortune*, 9 March 1981, 123.

2. William I. Strauss, ed., *Status '81: A Report on the Integrated Circuit Industry* (Scottsdale, Ariz.: Integrated Circuit Engineering Corporation, 1981), 10.

3. According to Glen Madland, president of Integrated Circuit Engineering, a

leading technical consulting firm, "At high densities, truly standard chips don't exist." See "The Boom in Tailor-Made Chips," 123.

4. Mel H. Eklund and William I. Strauss, eds., *Status '82: A Report on the Integrated Circuit Industry* (Scottsdale, Ariz.: Integrated Circuit Engineering Corporation, 1982).

5. Eklund and Strauss, *Status '82*, 9, 47–48; "The Boom in Tailor-Made Chips," 126.

6. Eklund and Strauss, *Status '82*, 47.

7. Ted K. Bradshaw, "The Future of the Electronics Industry in California" (Working paper, Institute of Government Studies, University of California, Berkeley, 1983), 1.

8. "Silicon Valley's Maternity Ward: How a Company Was Born," *New York Times*, 7 March 1982, F:4–5.

9. Eklund and Strauss, *Status '82*, 39.

10. Ernest Braun and Stuart MacDonald, *Revolution in Miniature: The History and Impact of Semiconductor Electronics* (Cambridge: Cambridge University Press, 1978), 124–25.

11. See, for example, David C. O'Connor, "Changing Patterns of International Production in the Semiconductor Industry: The Role of Transnational Corporations" (Paper prepared for the Conference on Microelectronics in Transition, University of California, Santa Cruz, 12–15 May 1983); and Linda Y. C. Lim, "Global Factory in the Global City: Labor and Location in the Evolution of the Electronics Industry in Singapore and Southeast Asia" (Working paper, Center for South and Southeast Asian Studies, University of Michigan, Ann Arbor, 1983).

12. See, for example, "Can Semiconductors Survive Big Business?" *Business Week*, 3 December 1979, 66–81.

13. Eklund and Strauss, *Status '82*, 49.

14. "Joining Hands against Japan," *Business Week*, 8 December 1980, 108–13; "Electronics Firms Plug into the Universities," *Science*, 6 August 1982, 511.

15. Arthur L. Robinson, "Giant Corporations from Tiny Chips Grow," *Science*, 2 May 1980, 482.

16. Arthur L. Robinson, "Are VLSI Microcircuits Too Hard to Design?" *Science*, 11 July 1980, 262.

17. A. Robinson, "Are VLSI Microcircuits Too Hard to Design?" 258.

18. Eklund and Strauss, *Status '82*, 33, 36.

19. Eklund and Strauss, *Status '82*, 36; Nico Hazewindus, *The Microelectronics Capability of the United States* (New York: New York University Center for Science and Technology Policy, 1981), 175–78.

4 Rethinking the Skill Requirements of New Technologies

Paul S. Adler

The impact of new technologies on skill requirements has always aroused controversy. It is instructive to compare our current debates with those of the previous wave of concern, the so-called "automation scare" of the 1950s and early 1960s.[1] In the 1950s, discussion was cast in terms quite different from those that predominate today. James Bright, one of the most perceptive participants in the earlier debate, summarized the key issue thus: "Not only will the average worker be displaced by the higher productivity of automatic equipment; he will be barred from the plant because he lacks the education, training, and skill necessary to hold one of the automated jobs." Bright's own intervention in the debate, however, ran against the stream. An in-depth survey of some thirteen firms in various industries led him to conclude that "automation had *reduced* the skill requirements of the operating work force, and occasionally of the entire factory including the maintenance organization."[2] A "de-skilling" prognosis similar to Bright's is the one most often heard in the automation debate of the 1980s.[3]

Widely different political views can be curiously allied in their pessimistic conviction that automation is "bad news" for the worker. Recent sociological research has given prominence to the work of radical critics, like Harry Braverman, who denounce capitalism's "secular trend toward the incessant lowering of the working class as a whole below its previous conditions of skill and labor."[4] Braverman and his colleagues see in Bright's research (indeed rely on it for) justification of their thesis that workers should seek to reshape a social system that progressively "de-skills" work. Bright himself, on the other hand, saw his results as comforting to managers. Automation, he argued, would generate little need for operator retraining and no overall shortage of suitable labor.

But Bright's diagnosis was faulty.[5] He was correct insofar as the "sudden massive upgrading" thesis is clearly untenable. But Bright overstated his case. Automation does correspond to an upgrading of skill requirements, albeit in a frustratingly slow and erratic manner.[6]

Job requirements do tend to increase, and perhaps more importantly, they change in form. It is because he missed the qualitative change in types of skill that Bright, like so many other observers, mistook the direction of the shift.

Bright built his survey around a relatively standard twelve-category breakdown of distinct "worker contributions," derived from the experience of job-evaluation specialists. This chapter analyzes each of these twelve contributions as they have evolved in the course of the computerization of clerical operations in banking. The principal result is that bank computerization has engendered several qualitative shifts in the nature of workers' contributions: (1) employee responsibility has come to play a more pervasive and decisive role in maintaining the quality and the efficiency of operations; (2) employees' intellectual grasp of the underlying banking and computer processes has become a critical necessity; and (3) their capacity for formal and informal cooperation has been highlighted by the growing interdependency of system components. These qualitative changes can be seen to have imposed themselves with some brutality, and with direct repercussions for both training requirements and the nature of work intensity.

The primary thesis explored here with the support of the banking example is that valid analysis of technologies' skill requirements must deal not only with the *quantity* of training workers need but also with the *qualities* for which they are to be trained. As highlighted by the case of bank employees, these qualities can be analyzed in three "dimensions": responsibility, abstractness, and interdependence. A second thesis, suggested but certainly not proven here, is that although local effects of automation are quite varied and are mediated to a considerable extent by a host of other factors, the underlying logic of automation tends to draw work in specific directions along these dimensions, toward broader responsibility, more abstractness, and higher forms of interdependence. Training times thereby tend to lengthen, and the content of training becomes more general and scientific, less narrow and specific. Work effort thus becomes more mental than physical, and more continual than continuous.

The consequences of this diagnosis for management are clear. In an economy in which advances in productivity are an imperative thrust upon firms by competitive pressure, inappropriate labor policies can be mortal. Many U.S. firms have benefited from reductions in labor cost and increases in process control derived from their de-skilling orientation toward blue- and white-collar operative jobs. But evidence from the banking industry confirms the results found by others in manufacturing: however advantageous in other respects, this de-skilling orientation cannot sustain long-run competitiveness in a technologically dynamic economy. An overly restrictive definition of the workers' role deprives them of the operating space and the technological culture necessary to deploy successively more advanced generations of equipment. Japanese and German firms' recognition of the key role of a continual, if not necessarily rapid, upgrading of their workers' capacities has given them an important competitive edge. This advantage can only increase as the pace of technological change quickens. It is worth noting, too, that this diagnosis applies not only to "high-technology" contexts; careful consideration of the three qualitative dimensions of work is critical for competitiveness in relatively mature and traditional industries, many of which are undergoing major, possibly "dematuring," technological changes.[7]

In order to capture the broader trends in skill, the analysis presented below is not concerned with the relatively few high-level jobs created by automation, but instead with precisely those relatively low-level, low-paying jobs that proponents of the pessimistic, de-skilling point of view like to highlight. It is, after all, in the name of these, the majority of workers, that such a respected scholar as David Noble has recently unfurled the banner of Luddism.[8] At stake, therefore, is whether automation as we know it is a force of social progress, and if so in what sense.

Automation in Banking

Let us first review the history of bank automation. Examples presented in this chapter are from French banking, with which I am more familiar.[9] Recent developments in banking in France have been perhaps even more interesting than those in the United States. Over the period from 1950 to 1980, the number of bank accounts in France increased by a factor of more than five, encouraging banks—which are, on average, monsters by U.S. standards[10]—to undertake very ambitious computerization programs. The general characteristics of both old and new systems are, however, quite comparable in the two countries. The history of bank automation offers an interesting example of the classic transition from the worker-dominated *tool*, to the worker-dominating *machine*, and then to the *automatic system*, whose relation to the worker is the object of this study.

Stage 1. In the least technically developed systems, predominant until at least the 1930s, where the means of production appeared as tools, the process flow ran in a linear fashion through the work sites.

The back office of the local bank branch sorted the paper traces of transactions effected in the front office, updated clients' accounts, and communicated results (in document form) to the head branch that handled regional accounts. The regional office would then transmit this synthesis to the national head office for the "fabrication" of aggregate accounts. Information retrieval was largely from locally maintained document files (the branch kept handwritten account cards) and was supplemented with a reverse flow of aggregate control information from central to peripheral sites. All sorting and accounting was accomplished manually. The available equipment—adding machines and typewriters—was used as tools, entirely subordinate to an individual worker's task.

Stage 2. As the mass of information grew with the extension of banking services in the interwar period, so did the specialization of the processing function. The branches, local and head, continued to handle most processing of their clients' accounts, but the bank progressively removed the fabrication of internal accounts from the branch to specialized central processing offices. The logic of this spatial and functional specialization was at first purely organizational; it held no significant technical or cost advantage. Most banks simply thought that branch personnel should concentrate on their "sales" role.

Beginning in the 1930s, mechanical business machines made their appearance; they were to gain predominance only in the 1950s. But both before and after mechanization, the sequence of processing operations remained basically unchanged from the linear form of stage 1, even if a somewhat greater proportion of local branch information was processed in and retrieved from other units. Work in the processing offices became dominated by the machine. An industrial type of work environment made its appearance in banking. In French banks the large concentration of employees, the nature of their work, and the low pay provoked a series of strikes in the processing centers in the late 1950s. For branch and head office employees, however, the mechanical business machines remained essentially tools that were used to prepare accounts more efficiently.

Stage 3. During the 1950s, even while mechanization was still being completed, banks began introducing computers. The continual progression of paperwork associated with the growth of clientele was forcing the pace of improvements in productivity. Data processing was in the "batch" mode, with overnight processing; the liaison between the computer center and the user departments involved card-punching (or, later, magnetic tape recording) for data entry, and the physical transportation of computer printouts. The specialized processing centers created in stage 2 changed tools and tasks, but the overall process flow was otherwise unchanged. To the outside observer, computers thus seemed mere extensions of the mechanical systems with which they coexisted. But a profound shift had taken place between stage 2, with or without mechanical business machines, and stage 3. In the later stage, the computer had taken over major portions of the process of fabricating accounts, the central activity of the bank. The sequence of operations previously conducted manually or with the aid of various machines was now *internal* to the computer system.

Stage 4. The next phase was characterized by the spread of computer terminals into all the units. First, they were installed in the processing and accounting units. Then, towards 1970, they appeared, in purely interrogative mode (data could be accessed but not entered), in the branches, where they offered up-to-date balance information. Beginning about 1978, interactive terminals allowing data entry as well as data access began to appear in all bank departments—a development that was largely complete in the major French banks by 1985. This new configuration represented a major shift made possible by the availability of cheap intelligent terminals. Management was quick to seize upon this potential, especially as labor tension grew in the processing workshops. These tensions were manifested in France by a major, crippling strike in 1974; in the United States they appeared in the form of exacerbated turnover rates.[11] In both countries the specialization of data-entry tasks also generated intolerably high error rates.

French banks thus abandoned the practice of keeping productive functions centralized and separated from commercial functions. The organizational principle became to despecialize data entry as far as possible, making each employee responsible for the accuracy of his or her own data. Processing was progressively decentralized out to the back office of the local branch and even right out to the front desk. (Automatic Teller Machines take this decentralization to its logical conclusion by making data entry the customer's task.) When entering data, the teller, like other users, is now automatically informed of any inconsistencies or implausible data that the system has detected, and is asked to rectify the entry. Thus processing time and errors are minimized. With account fabrication now automated within the computer system itself, the remaining specialized processing centers are reduced to the treatment of anomalies and residues. Tasks centralized in the head office are also transformed; the administrative role there now accentuates research and control.

At stage 4, the current phase, the full implications of stage 3 become apparent: the role of the worker is to assure the interface between the computer and the client. The parallels with the high degree of automation found in Bright's petrochemical refinery case are striking.[12] Long sequences of operations are conducted without any human intervention, except for the input of raw materials and a (more or less) constant surveillance. The system monitors the quality of inputs (in banking, information) and the compatibility of its own suboperations (by account-balancing tests) and, if sufficiently advanced, as with time-sharing on large-scale on-line computer-terminal systems, "anticipates action required and adjusts to provide it"—Bright's characterization of the highest levels of automation.[13]

An Example: The Transformation of Demand Deposit Accounting

This four-stage evolution has revolutionized the labor process in banking. With the passage from single-work-station machines to technical systems embracing large collectivities, the worker has become increasingly peripheral to the basic operation of account fabrication. This transformation can be seen perhaps most clearly if, following Bright's example, one constructs a "mechanization profile," charting on one axis the sequence of basic tasks and on the other axis the level of mechanization of each component task.[14] It is especially instructive to do this for the most labor-consuming of bank operations, demand deposit accounting.[15]

Figure 1 can be taken as a condensed description of the "best practice" available in the four stages of bank technology presented in the preceding section: (1) manual, (2) mechanical, (3) batch computer processing, and (4) on-line computer processing.[16] The figure makes clear the extent to which lower and earlier phases of mechanization span only a limited number of operations, leaving machine-aided tasks separated by manual operations. As banking moves from mechanization (stage 2) to computerization (stages 3 and 4), a series of tasks formerly considered the very essence of bank work have been eliminated, including accounting imputation and adjustment, classification of documents, multiple entries of data, manual data search, and supervision by signature. The work situation has been totally transformed.

The worker is now entirely dependent on the computer system—whose malfunctioning is frequent enough to constitute in itself a permanent feature of the computerized work situation. Moving into stage 3, and even more in stage 4, work activity is mediated by a new language of computer codes. A new range of tasks has been introduced.

Figure 1. Stages of demand deposit accounts processing

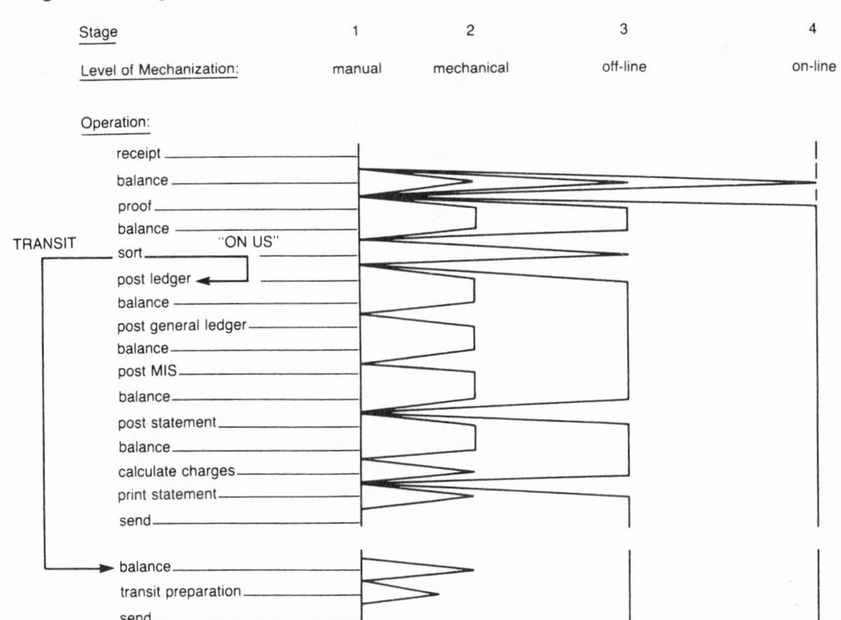

Note: Checks that are "in transit" are checks not drawn "on us" but on another bank. Their processing is truncated.

Accountants now diagnose and rectify residues and anomalies listed by the computer system. New types of errors—and new types of fraud—appear. Their greater complexity, more important consequences, greater cost, and more difficult diagnosis are congruent with the level of automation. Surveillance of work no longer takes the form of a document-by-document verification and the signature by officers of large-sum transaction records. It has instead become a combination of computerized control of the consistency and plausibility of data, verification of users' personal identification codes, and the supervisor's en masse, ex post surveillance of a computer record of extraordinary operations.[17]

In the refineries Bright surveyed, production was a continuous-flow process. In banks we witness a literally instantaneous production process. Once data (the raw material) enter the bank's records in any form at any place, they are automatically and instantaneously fed into all the pertinent accounts including the client's account, the bank's general ledger, and its management information system, auxiliary accounts, official accounts, and control accounts.[18]

A general characteristic of higher levels of automation emerges. Reducing repeated entry of the same data and controlling automatically for data plausibility both permit a major reduction in the *frequency* of errors. *Total cost* of errors will normally be reduced, but errors that do persist and the new sorts of error proper to such systems have a much greater *unit cost*, due to their instantaneous propagation, their inherent complexity, their unpredictable form, and correspondingly increased costs of discovery and rectification. The nature of the quality-control problem is transformed.

Note that demand deposit accounting is the worst case for my thesis. It was in precisely this area of bank operations that data-entry functions gave rise to the largest mass of very low-skill jobs. But this specialization characterizes stages 2 and 3, not stage 4. As the preceding historical sketch shows, the requirements of productivity and quality, when confronted by new technological possibilities and by the labor-relations cost of de-skilling, overcome the attraction that ultra-simplification of job design may hold for management.

Automation and Skill Requirements

Bright's characterization of the effects of increased automation on skill requirements in the firms he surveyed is presented in table 1. His scheme provides a useful structure for reviewing developments in the banking industry. However, two introductory comments are necessary.

First, Bright's table, like my analysis, refers only to operators. As he pointed out, the proportion of operators to other personnel, like maintenance and technicians, will vary with the level of automation.[19] He is emphatic that the number of maintenance people shows no general tendency to rise with the level of automation; but maintenance staff's skill requirements do change significantly and in general rise. This means that even if we accept Bright's characterization of the effects of higher levels of automation on operators' skill requirements, it does not follow that average skill levels of the *whole* labor force will likewise decline.

Second, Bright was above all concerned with the passage from the second column of the table to the third. His study of the highest category was limited to the case of a single refinery—where, moreover, the control-room function was reserved for a supervisor. Not surprisingly, Bright's analysis of skill requirements at the highest automation levels is sketchy. His main focus was appropriate to the spectrum of

Table 1. Bright's Analysis of the Changing Contribution Required of Operators with Advances in Levels of Automation

WORKER CONTRIBUTION * OR SACRIFICE TRADITIONALLY RECEIVING COMPENSATION	MECHANIZATION LEVELS			
	HAND CONTROL	MECHANICAL CONTROL	VARIABLE CONTROL, SIGNAL RESPONSE	VARIABLE CONTROL, ACTION RESPONSE
PHYSICAL EFFORT	INCREASING-DECREASING	DECREASING	DECREASING-NIL	NIL
MENTAL EFFORT	INCREASING	INCREASING-DECREASING	INCREASING OR DECREASING	DECREASING-NIL
MANIPULATIVE SKILL (DEXTERITY)	INCREASING	DECREASING	DECREASING-NIL	NIL
GENERAL SKILL	INCREASING	INCREASING	INCREASING-DECREASING	DECREASING-NIL
EDUCATION	INCREASING	INCREASING	INCREASING OR DECREASING	INCREASING OR DECREASING
EXPERIENCE	INCREASING	INCREASING-DECREASING	INCREASING-DECREASING	DECREASING-NIL
EXPOSURE TO HAZARDS	INCREASING	DECREASING	DECREASING	NIL
ACCEPTANCE OF UNDESIRABLE JOB CONDITIONS	INCREASING	DECREASING	DECREASING-NIL	DECREASING-NIL
RESPONSIBILITY †	INCREASING	INCREASING	INCREASING-DECREASING	INCREASING, DECREASING, OR NIL
DECISION MAKING	INCREASING	INCREASING-DECREASING	DECREASING	DECREASING-NIL
INFLUENCE ON PRODUCTIVITY ‡	INCREASING	INCREASING-DECREASING OR NIL	DECREASING-NIL	NIL
SENIORITY	NOT AFFECTED	NOT AFFECTED	NOT AFFECTED	NOT AFFECTED

* Refers to operators and not to setup men, maintenance men, engineers, or supervisors.
† Safety of equipment, of the product, of other people.
‡ Refers to opportunity for the worker to increase output through extra effort, skill, or judgment.

Source: James R. Bright, "Does Automation Raise Skill Requirements?" Harvard Business Review 36, no. 4 (July-August 1958): 92. Reproduced with permission.

technologies prevailing in the 1950s, but inadequate in view of more recent developments.

Let us now review each of Bright's worker contribution factors, comparing his diagnosis with the banks' experience.[20]

Physical effort. Bright's assessment that physical effort requirements decline with increased automation seems difficult to dispute. He underestimated various ergonomic problems—which, it is true, have only recently come into prominence. It is unclear under which rubric such problems should be classed; here they have been placed in the section on hazards (see below).

Mental effort. Bright's diagnosis that increased automation requires reduced mental effort reveals an important shortcoming in the application of his framework. In his view, at the highest levels of automation, "by definition, the more automatic machines employ control devices that regulate their performance to achieve the desired end *without human attention*. Therefore, mental strain as a result of mental effort is ultimately reduced."[21] The automatic control capacity of an advanced machine system may well reduce the human role *quantita-*

tively; such is the basis for Bright's argument. But he seems to have been so impressed with automation's potential for self-regulation that he ignored the original question: given that employment levels will adjust to reduce any redundancies in the work force, what is the *qualitative* nature of the remaining jobs?[22]

In banking, increasing automation has meant the replacement of a sequence of elementary data entry tasks by a system in which a single operator enters into the system the variables needed for each transaction. A simple operation like cashing a check involves nineteen distinct tasks of several different kinds (data entry, data verification, terminal procedural commands, physical manipulation) that the teller, working with an on-line terminal, must execute in precise order, respecting the software's sequence of decision branches. Each separate task is simple, but their correct combination and ordering, especially while facing a line of impatient customers, demands the internalization of a new sort of intellectual discipline: the ability to master procedural, algorithmic modes of thought.[23] The clerical worker's task is to decide, as a function of the situation with which he or she is confronted, which procedure corresponds to the task, how to correlate elements of the situation with elements of the procedure, how to deal with any abnormal, exceptional elements of the situation, and how to manipulate the tools with which to effect the procedure. The impact of this aspect of automation on skill requirements receives further discussion below. It should simply be noted here that the mental effort involved in constantly maintaining a "clear head" is not to be underestimated. It is a new form of effort for many workers.

Another feature of particular importance is that interaction with the automatic system (as with advanced manufacturing equipment) is mediated by a multiplicity of codes. Even when software becomes more user-friendly, the operator–machine interface is not one characterized by truly "natural" language. A common illusion here is that habit can make the new code-mediated tasks banal. Detailed research has shown that even after a long apprenticeship, coded data-entry involves perceptual processes that are ergonomically more demanding than those involved in the entry of regular text—whence a second source of mental effort that Bright ignores.[24]

Outside the sphere of low-level bank operations, there is another, less obvious, form of mental effort that Bright underestimates. In discussing the highest automation levels, he wrote: "The progressive effect of automation is first to relieve the operator of manual effort and then to relieve him of the need to apply continuous mental effort."[25] He described the role of the operator of more advanced machinery as "patrolling" or "machine-tending," evoking the image of the night watchman. Subsequent studies of refinery workers have shown, however, that the mental effort of surveillance imposes a considerable burden of a new kind: the *continuous* concentration associated with fabrication is replaced by the strain of *continual* watchfulness and readiness.[26] Highly automated systems generate a new and complex operating mode that demands of workers that they be ready to react rapidly even after hours of apparent inactivity. Associated with this new operating mode is a new sort of strain due to the conjunction of (1) high levels of responsibility (see below) and (2) modes of system failure increasingly difficult to predict.[27]

Bright ignored problems such as these in his faith that machines would eventually control themselves. But as control functions become more sophisticated, the problem is only removed to higher levels of responsibility.

Manipulative skill (dexterity). As with physical effort, it is clearly difficult to dispute Bright's claim that dexterity requirements decline as automation levels rise. Typing skills will become a more common requirement, but in their specifically manual component these are of a lower level of complexity than many of the manual skills automation displaces. But consideration of this factor should not be limited by its implicit point of reference—the manual use of simple tools. Beyond simple dexterity, there is a whole range of "tricks of the trade" that constitute the "art" of many occupations, white- and blue-collar alike. These are in large measure the fruit of experience, and are discussed under that heading below.

General skill. In classical economic theory, relative wages fluctuate around a value determined by the "cost of producing" competence that is embodied in training (formal education, vocational training, on-the-job training, and experience). It is therefore of particular interest to find that the case of banking runs absolutely counter to Bright's conclusion that general skill requirements fall with increasing automation.

First, workers must learn a new language. It is quite wrong to believe that operators must merely learn a few simple codes by rote. Some commentators seem to believe that just because the operations have been codified in a manual, they no longer require any skill, as if the operator simply follows instructions. One need only think of the airline pilot's job and its exhaustive documentation to realize that the crucial variable remains the operator's "internalization" of the instructions. As code-mediated operations become more pervasive, the nature of the work to which workers must adapt themselves is transformed. And learning a new language is never a trivial adaptation.

Management in French banks has been tempted to limit the extent of this new literacy. Training programs are expensive and can be cited as justification for higher pay scales. But the managers I spoke to expressed their concern with another, countervailing consideration: that such a cost-focused human resource management policy would prove to be myopic, not only by limiting operators' capability to fully master their current tasks, but also, and perhaps more importantly, by hampering their ability to adapt to constantly evolving computer procedures. In the short run, it may seem efficient to separate routine tasks requiring little thought (even if they require more mental effort than is commonly acknowledged) from the treatment of exceptional cases—which can then be reserved for supervisory staff. In the longer run, however, this strategy leads to burgeoning overhead costs and a decline in the quality of work; neither outcome is socially or technically unavoidable, and both are serious handicaps in the competitive struggle. In the longer run, successful operations management dictates that workers be given the resources they need in order to do their work—without calling on supervisors every time the slightest mishap occurs. These considerations are leading French bank management to develop computer literacy programs for their employees. They are also active in encouraging the Education Ministry to adopt high school programs that would help them to avoid this (direct) expense.

The costs of inadequate training are not only felt by management, as loss of operations efficiency, but are also borne by workers, in the form of strain. Ergonomists have observed that mental strain is generated by overly restricting workers' understanding of the language they are using.[28] This strain is another element to be considered in assessing the mental effort required in automated work (see above).

Bright's study ignored such changes in the operating "tools" of the worker. A second flaw in his assessment lies in his characterization of operating "tasks." He wrote: "In many instances, the need for education and understanding of principles may continue well into the higher levels [of automation]. However, these eventually become unnecessary contributions as reliability increases."[29] So imagined the bank managers I interviewed—*before* the computerization process moved into the highest levels of automation.

The managers responsible for the stage 4 branch computerization that is now under way were haunted by the following scenario: the client comes in, angry, wanting to know why the money transfer he/she effected a month ago has not been credited to his/her account; the teller has no one to turn to—indeed it was the teller who had effected the operation on the terminal at the front desk. There is no longer a chain of processing offices handling such operations, and hence there is no longer a series of office chiefs to whom the teller can turn in order to determine where along the processing chain the operation has broken down. Thus the "bankers' paradox": they had indeed hoped to simplify operations and therefore reduce training requirements when they computerized. Behind the public presentations about "enriched teller operations," such was in fact their main concern. And now, committed to a vast program of branch computerization, they were discovering—with no little anxiety—the magnitude of the training task before them. If a modicum of what they have come to call "local mastery" were not sustained at the level of the local branch and the front-desk teller, the bank was liable to find itself swamped by angry customers. The lesson from the banking case is straightforward: the need for "local mastery" grows as the level of automation increases.

The only possible response to this pressing need is training. In banking, three areas of training development have emerged as particularly important: (1) basic computer literacy; (2) information regarding the structure of the bank's processing system; and (3) for item 2 to have any meaning for the employees, a much broader understanding of the logic of bank accounting. The bank I studied most closely, like others in France, was therefore busy developing educational courses, software, and video cassettes in these three areas. It was still unclear how many hours of training would be necessary. But because of their clients' expectations of accuracy, banks were not, unlike many other industries, able to bury their heads in the sand and ignore the training imperative associated with all but the technically most elementary jobs: training must go beyond the immediately functional if it is to be really operational.

As in Bright's refinery case, banking labor productivity increases with automation. Fewer workers are necessary to accomplish the same number of transactions. Their training in rote learning of complex banking practices is perhaps progressively less important. But rote learning is replaced by cognitive learning in computer basics and the technological and accounting logic of transaction processing.

A crucial corollary arises from this analysis of operator skill requirements: the level of automation represented by these banking systems permits great flexibility in reprogramming and in the proliferation of special-purpose applications. The common assumption that higher levels of automation mean greater rigidity of process no longer holds.[30] As a result, the interdependence of Development and Operations moves to a qualitatively higher plane and becomes a critical competitive factor. The implications for human resource management policy merit serious consideration.

General education. Clearly, sooner or later some of the technological training described above under "general skill" will be accomplished in the framework of general education. One could even hypothesize that an increasing proportion of training requirements will thus become common to an increasingly homogeneous set of tasks in various industries and occupations. The "skill" hierarchy could thereby be eventually flattened, not by de-skilling, but by virtue of a general rise in the training level of less-skilled labor. But to restrict discussion to recent debates—which can be heard in all advanced industrial countries—it is apparent that a broad program of technological literacy is an urgent requirement as automation spreads in modern societies. The problem is not merely a matter of particularly acute bottlenecks with engineering personnel. At stake is the capacity of broad sectors of the labor force to deploy automated, and especially computerized, systems effectively, in the office as in the plant, not to mention in the home.

Furthermore, it is not simply the quantity of education demanded that is changing, but also its content and form. The "clear thinking" requirements of procedural operating modes seem to pose a challenge to the educational system. We are witnessing a cultural mutation of considerable proportions, perhaps to be compared to the "Americanization" experienced by the wave of immigrants that populated the Ford assembly line at the beginning of this century. Ford's objective was to develop body disciplines and life-styles that would permit immigrants from largely rural backgrounds to operate assembly-line facilities. Today, "educators are starting to ask whether the way Americans think, the way they go about defining and solving problems, will be altered, for better or for worse, by increasing experience with the systematic and quantitative thinking that goes into programming a computer."[31]

Experience. Bright's thesis that increased automation brings declining experience requirements was dictated by his assumption that the machine will eventually be totally reliable in doing what the worker once did. I assume my point on this score has already been made. There is a further question, however, of considerable interest, that of the changing *relative* weights of education and experience requirements as the level of automation rises. To the extent that skill requirements of literacy, numeracy, and technological culture rise, one could hypothesize that not only dexterity but, more generally, the whole range of "tricks of the trade" will play a correspondingly reduced role in workers' contributions.

Banking presents another lesson on this question. The need for basic accounting training for low-level bank personnel has already been emphasized (under general skill, above). Pushed as they are to the periphery of the account fabrication process, their mastery of their own tasks demands some understanding not just of the computer system in its technical dimension, but also of the accounting logic that governs the nesting of bank accounts. In the previous stages of banking technique, no such breadth of training was required; *all* the functional training for low-level positions was acquired on the job. This no longer suffices. The problem becomes even more acute with positions requiring greater technical expertise and/or responsibility. Previously, it was by successive apprenticeships in several departments that upwardly mobile personnel learned their banking. Now experience cannot suffice, as critical operating procedures have been internalized in the computer system. Thus in both instances, for operating and for supervisory personnel, training requirements are increased and become more formal and theoretical, less based in on-the-job experience.

This shift in the form of training reinforces the importance of general education relative to that of experience. However, the roles of the various training agencies—schools, industry-level institutes, firm-level training departments—are still to be determined. Remember that computer systems, despite being situated at very high levels of automation, are not entirely standardized—far from it. Moreover, while small-scale users usually buy packaged software, larger users are constantly adapting their in-house applications. The result is that many operators have to learn surprisingly idiosyncratic and constantly evolving company-specific systems. This diversity may have the effect of reducing the transferability of skills from one firm to another. The relative accessibility of different kinds of training agencies will reflect and influence the ability of firms to entrap workers by making their skills more firm-specific.

But such trends, of course, cut both ways. If workers find it difficult to sell their skills on the external labor market, management is by the same token forced to acknowledge the greater value of stable workers. Turnover becomes more expensive because of the duration of training and because of the greater cost of the disruption of formal or informal work teams. The decisive feature of this situation is that the terms of the trade-off between stability and instability of employment are transformed. However firm-specific systems may be, workers need to be able to adapt to their continual evolution, and they will be all the better equipped for this learning when they possess a solid computer literacy.

At a more general level, the terms of the trade-off between specificity and generality in training are similarly transformed. Automation encourages an overall minimization of the importance of "tricks of the trade," to the benefit of more general, and therefore more generalizable, knowledge; skills become correspondingly less occupation-specific. Evidence from the experience of more automated firms is also encouraging, in that job switching and flexibility in work assignments seem to increase with the level of automation.[32]

Exposure to hazards. It seems reasonable to hypothesize that as workers become less directly involved in product manipulation, hazards in the workplace will probably, as a general rule, decline. It is likewise understandable that as the harsh conditions of yesteryear are forgotten, the relatively minor inconveniences of today are easily magnified. But some ergonomic problems associated with current computerized equipment cannot easily be relegated to the status of "undesirable job conditions."

Computer display systems can aggravate eye and back strain. The problem is partly in the design of the equipment and its installation. Too often no provision is made for adjustment of lighting, height, tilt, physical configuration of screen and keyboard, and so forth, in keeping with elementary industrial engineering standards. Another part of the problem resides in the current technology of cathode-ray tubes, whose deficiencies include limited definition, high scintillation, and inadequate contrast control.[33] Clearly improvements in this area are needed. Unrectified, such strain can lead to serious injury. Equally clearly, these problems are not inherent in the current level of automation; they are surmountable, if only under the impetus of organized intervention. Some French banks are already using liquid crystal displays, which dramatically reduce some of these problems.

The key ergonomic problem, however, seems to lie not with the technology, but with the organization of work. High-concentration tasks with highly constrained operating modes, exemplified in jobs such as those of data-entry clerks, should, for ergonomic reasons, not be maintained for more than a couple of hours at a time.[34] This has been the recommendation of various European studies, and more and more employers are responding to the difficulty by despecialization, incorporating data-entry tasks into other jobs, like the bank teller's.

Thus without minimizing equipment-related hazards, it can be said that not a few of the ergonomic problems ascribed to new technologies reflect a displacement onto technology of frustration with other factors, like work organization or job content. We shall see that the very nature of automated work may be aggravating these other factors.

Undesirable job conditions. Effort requirements in the banking industry may be on the decline, skill requirements seem to be on the increase; yet jobs are not experienced as any more satisfying. On the contrary, the humdrum nature of many banking jobs is eventually classified as an "undesirable job condition" in itself. For example, French banks have been forced to give a high priority to automating securities department processing jobs, for the simple reason that these jobs were experienced as so boring that stable staffing could not be assured.

One can thus venture a simple hypothesis that goes beyond the banking case: as the fabrication process is automated, and as the worker is rendered increasingly peripheral to it, the operator's work is reduced from its concrete, tangible form to a more abstract activity—interface and surveillance. In this transition, the concreteness of the task's goal is diminished. In place of the easily apprehended goal of correctly effecting a distinct operation, the new situation substitutes a much more amorphous objective, that of ensuring the smooth functioning of an almost totally automated set of interrelated operations. That is, not only the *means*—computer-mediated transactions via coded commands—but also the *ends* of work appear as less concrete and more abstract.

As discussed earlier, the abstraction of means leads to increased training requirements and new kinds of mental effort. The abstraction of ends leads to a very different result: work is often experienced as boring. In the processing departments of French banks this lack of identifiable tangible objectives led to such high error rates that management responded by making each clerk responsible for a group of branches. In the United States, Citibank has undertaken a similar reorganization. Despite the operational value of these "responsibilization" strategies, such efforts appear to be attempts to accommodate an underlying irreversible trend. Far from restoring the craftwork type of job satisfaction, they substitute for it something far closer to a disciplinary constraint. It is difficult for operators to avoid the impression that it is the errors they are noted for, not the quality of their effort. At the front desk, on the other hand, such "responsibilization" can serve as a more viable job-enrichment factor and can offer the customer a more broadly competent interlocutor.

Obviously reorganization of work can do much to alleviate an underlying loss of inherent interest. But the front-desk example is not easily generalized, for its success derives from the partial substitution of a sales objective for a technical interface objective. As automation progressively pushes workers to the periphery of production, and to the extent that production tasks are not supplanted by sales tasks, the abstraction of ends seems to exacerbate the old problem of boredom with work. Can we not therefore predict that over the longer run will come a growing demand for reduced working hours? There is clearly a segment of the work force—programmers, analysts, and the like—for whom the "concrete object" of their efforts is the system itself. Their case is more complex. But for the mass of workers who perform the interface and surveillance functions, like bank employees in their vast majority, only three factors seem to limit the appeal of a reduced work-week: income needs, the social value of the work community, and the perceived limits of much extra-work life. As Mark Twain put it (in another epoch): "I'm glad to be a man; it gives me somewhere to go in the morning."

Responsibility. Despite Bright's doubts, the evidence from banking, as well as a closer reading of his own refinery case, would indicate fairly unambiguously that responsibility generally increases with automation. The case of the refinery control-room worker is perfectly clear. It is only the assumption that the systems will run themselves that led Bright to doubt the longer-term evolution of the responsibility requirements of this job. But of course it is difficult to imagine such control-room jobs spreading to encompass many industries in the foreseeable future.

In this context the banking case is revealing. The key consideration here is the multiplicity of operations that the computer conducts on the basis of a single data entry: the teller's operation, sometimes augmented by that of the remaining processing units, feeds a multiplicity of accounts at all levels of the bank's records. The banks have therefore invested considerable effort in developing checks for the consistency and plausibility of data. These, however, are never perfect. The less frequent errors become, the greater the complexity and the unit cost of the remaining errors.

Bank managers' interest in maintaining "local mastery" is rooted in such problems. Similarly, the reorganization of the remaining processing departments into zones of geographic responsibility bears witness to the operational motivation for this concern with responsibility (beyond the need, described above, to alleviate boredom and reduce employee turnover). The import of such an increase in responsibility, even for banal banking jobs, is, I believe, considerable.

The central issue is that as automation pushes workers out of the center of the production process, the responsibility solicited from them is no longer confined to the object of maintaining work-effort standards. Workers are asked to play a different role: to be responsible for results. Of course, managers have always wanted as much. What has changed is that now the technical characteristics of production make such responsibility an operational imperative. This entails a number of complications.

First, in the determination of wages, the responsibility factor seems to play a destabilizing role. Of all the worker contributions discussed thus far, responsibility is the most difficult to evaluate. In the dispassionate gaze of the academic, such may not seem the case, but amid the social tensions of a real-life firm it is more difficult to avoid charges of "subjectiveness." Clearly, neither supply and demand nor training time can be used as a means of evaluating the "sense of responsibility" of the worker. If, on the other hand, it is the job, rather than the worker, that is being evaluated, as in standard job-evaluation systems, conflict is often generated when it appears that the responsibility factor is being used as a means of increasing relative rates of pay for supervisors.

A second complication is that responsibility often cannot be assigned to individual jobs. Indeed, although many counterexamples come to mind, it could be argued (as suggested above) that automation will encourage this interdependence, making operations an increasingly collective, rather than individual, affair.[35] But there seems to be some tension between the wage as a fundamentally individual incentive and efforts to reward the "teamwork" component of responsibility. Furthermore, team bonuses pose the same problems of nontransferability as firm-specific skills.

Finally, it should be recalled that increased worker responsibility has always been resisted by a sizable sector of the business community as a dangerous threat. The difficulty to which these skeptics point can be expressed with brutal simplicity: "Give them a finger and they'll want a hand." This attitude should not be lightly dismissed as obsolete or irrelevant. Clearly, some managers in some industrial relations contexts let such fears overwhelm them, exaggerating the workers' desire to "take over." It is nevertheless intriguing to speculate whether, over the longer run, the tendency of automation to call increasingly on workers' responsibility will not encourage demands by workers that will encroach on traditional management prerogatives. A general upward drift in educational requirements would perhaps have a similar effect. For now, it is clear that "responsibilization" strategies of operations management touch on delicate, but perhaps unavoidable problems of legitimation.

Decision making. Decision making is closely linked to responsibility. The difficulty of any diagnosis here is that the term conflates changes in the scope of decisions with changes in the degree of authority over a given scope. Leaving aside the former problem, the banking case lends some support to the thesis that authority can be centralized in a set of operating procedures written into the software, thereby reducing the autonomy of decision among local personnel. Such centralization, however, becomes costly—for example, in poor-quality results—when it undercuts the real mastery of local operations. Centralization of key (strategic) decisions is thus commonly combined with a change in the scope of decisions, to encourage a concomitant decentralization of minor (operating) decisions.

As others have also pointed out, a significant feature of the new configuration is that structures of organizational coordination themselves become more "abstract" as they are increasingly embodied in software routines rather than in the personal authority of the local management.[36] This automation of coordination has some dramatic repercussions on the supervisor's role. Technological trends seem to encourage a shift in the sources of a supervisor's legitimacy from technical expertise to a role as team facilitator.[37]

To conclude, however, as Bright did, that decision-making contributions of operatives are decreasing or nil, is manifestly inadequate.

Influence on productivity. The case need not be restated here why worker performance in the automated environment is decisive for operational efficiency. Bright's position does, however, reflect more profound features of the situation that merit meditation. Under automated conditions, major increases in productivity can no longer come from rationalizing the worker's gestures. Freeing the production process from its dependence on the craftman's dexterity and eliminating the technical inadequacies that created numerous "pores" in the working day have both clearly made possible a certain intensification of work effort. But meanwhile, behind the scenes as it were, the progressive mechanization and automation of production, and the economic rationality and operations rationale that have developed with it, have been working to attribute to human intervention a different type of efficacy. Science and technology progressively displace both dexterity and intensity of effort as the key to increased productivity. This tendency confirms the new, enhanced significance of the worker's role in guaranteeing the smooth functioning of the highly automated system.

Seniority. Seniority is often a surrogate for experience. As such, its importance is subject to conflicting pressures and may, as a result, be declining. A further downward pressure on the importance of seniority, perhaps transitory but currently causing significant hardship, is the belief that younger workers have greater flexibility than older ones.

Older workers are less likely to have the requisite new skills, and all too often they are presumed—often incorrectly—to be less able to adapt. The hidden assumption behind this prejudice is frequently that automation will reduce skill requirements and that "older" (that is, current) workers will hence be overqualified.

Conclusion

To synthesize this twelve-part analysis, Bright's evaluation of the impact of automation on worker contributions can be compared with the preceding observations on the automation of banking. Table 2 recapitulates the salient features of both evaluations, focusing on the passage from stage 3 to stage 4. Examination of this table leads to three principle conclusions.

First, in contrast with Bright's assessment, the banking case shows as many quantitative increases as decreases. And on some of the contributions that have assumed particular importance in banking, either Bright was undecided (general education, responsibility) or our examples contradict his thesis (mental effort, general skill, influence on productivity).

Second, and more important perhaps, is that Bright's "pessimism" is belied by changes in the relative weights of different worker contributions. Indeed most of the decreases observed can only be judged welcome: the decrease in major hazards; the substitution of mental for physical effort, and of general skill and education for manipulative skill, experience, and seniority.

Third, and most critically, the qualitative nature of many of the contributions has changed. Review of the last column of table 2 identifies the symptoms of a misfit between Bright's twelve-dimensional analysis and the observed evolution of work. The explanation seems to be that beneath Bright's structure there is a deeper one—which is summarized here in table 3. To the extent that an accurate assessment of these qualitative changes is a key to higher productivity, management would be taking a dangerously short-term perspective if it adopted a labor policy inspired by Bright's prognosis. There are, of course, technical and market configurations that call for low-cost, minimally trained labor. My analysis of banking, however, strongly suggests that these will, in general—and even aside from humanistic concerns for improved conditions in the workplace—constitute the exception, not the rule, in technologically dynamic firms.

The framework proposed in table 3 can be helpful not only in the

Table 2. Changing Contribution Required of Employees at the Highest Levels of Automation: Bright and the Banking Case Compared

	Worker Contribution	Bright's Assessment
1	Physical effort	Nil
2	Mental effort	Decreasing-Nil
3	Manipulative skill	Nil
4	General skill	Decreasing-Nil
5	General education	Increasing or Decreasing
6	Experience	Decreasing-Nil
7	Exposure to hazards	Nil
8	Acceptance of undesirable job conditions	Decreasing-Nil
9	Responsibility	Increasing, Decreasing, or Nil
10	Decision making	Decreasing or Nil
11	Influence on productivity	Nil
12	Seniority	Not affected

THE BANKING CASE

Quantitative Change	Qualitative Change
Decreasing	But see Hazards (strain)
Increasing	Algorithmic operating modes displace manual gesture
Decreasing	But see Experience ("tricks of the trade")
Increasing	Technological literacy; local mastery; cognitive learning
Increasing	Technological culture; intellectual discipline ("clear head")
Decreasing and/or Increasing	Decreasing: relative to education Increasing: dynamic large-system idiosyncrasy, value of stability and teamwork
Decreasing	But pervasive strain with highly constrained operating modes
Decreasing objectively, but Increasing subjectively	Intrinsic boredom displaces physical conditions
Increasing	Responsibility for results, not just effort
Decreasing and/or Increasing	Decreasing: centralization of parameters Increasing: decentralization of operating variables (Note: more abstract decision processes)
Increasing	Cost of errors and down time (work efficacy, not intensity)
Decreasing	See Experience

Table 3. The Underlying Qualitative Dimensions of Work

Issues	Dimensions	
• local mastery • influence on productivity	SOCIAL	sense of responsibility for results of production
	versus	
	PRIVATE	purely instrumentalist attitude to work, as in Taylor's "trained gorilla," responsible primarily for work effort
• algorithmic operating modes • cognitive learning • decision-making structure • new ergonomic problems • computer literacy • importance of general training	ABSTRACT	abstract, mental, machine-mediated tasks with cognitive learning but low inherent interest
	versus	
	CONCRETE	tangible immediacy of tasks and goals, with manual or rote learning
• cost of errors • integration and flexibility of large systems	COLLECTIVE	systemic interdependence; critical role of coordination between tasks, and between R&D and operations; job-switching and teamwork encouraged
	versus	
	INDIVIDUAL	stand-alone or purely mechanical, sequential linkage of individualized jobs

practical task of defining appropriate policy, but also in the larger task of assessing technology's impact on civilization. To draw out but one part of the answer to the question posed at the end of the introduction to this chapter—whether automation is a source of social progress, and if so how—we can refer to two of the qualitative dimensions of our analysis: the "responsibility" and "abstractness" of tasks.

The responsibility factor, so critical to the deployment of automation's massive and rapid productive capacity, introduces a *challenge* into the work process. The worker can no longer be conceptualized, as Frederick Taylor proposed, as a "trained gorilla" whose role is restricted to that of following Scientific Management's "One Best Way" instructions. Successful deployment of advanced technologies requires that motivated performance replace reluctant conformance. Workers are solicited not only by management, but also by the technical characteristics of the operation, to evidence a sense of responsibility for the success of the whole process. They are, in this sense, solicited in their "social" substance, no longer just in their capacity as "homo economicus."

Simultaneously, the process of production becomes, from the workers' point of view, a more abstract affair. Their skills must change in form—and this despite the fact the inherent *interest* of the work may be receding. There is a world of difference in the type and degree of interest that shoemaking can hold for, on the one hand, the traditional craftsman and, on the other, the control-room operator of a plastics plant responsible for the continuous, automated production of thousands of plastic sandals every hour. The knowledge required of the control-room operator may be greater; the responsibility certainly is. But "button-pushing" tasks make for very aggregate and mediate goals. It becomes increasingly difficult for the worker to identify personal goals with the operational objectives of the tasks at hand.

If profoundly positive effects on work can be attributed to technical progress, it must also be acknowledged that there is an important tension between these simultaneous tendencies towards greater challenge and responsibility and greater boredom and abstraction. Perhaps the debate on the social effects of our increasingly technology-intensive development has been too polarized into pro- and anti-technology camps. The habitual association of a job's "skill" and its "goodness" may be losing pertinence. Traditional concepts of skill, especially models based on the ideal of the craftsman, cannot capture the full ambiguity of a growing proportion of jobs that are simultaneously more challenging *and* more boring. The resolution of this ambiguity is, at this stage of the development of civilization, far from obvious. The future, in more ways than one, remains to be invented.

Notes

1. For a cross-section of views, see *Technology and the American Economy,* (Washington, D.C.: Government Printing Office, 1966).
2. James R. Bright, "Does Automation Raise Skill Requirements?" *Harvard*

Business Review 36, no. 4 (July–August 1958): 86 (emphasis in original); see also idem, *Automation and Management* (Boston: Harvard University Graduate School of Business Administration, 1958).

3. See, for example, Henry Levin and Russell Rumberger, "The Low Skill Future of High-Tech," *Technology Review* 86, no. 6 (August–September 1983): 18–21.

4. Harry Braverman, *Labor and Monopoly Capital: The Degradation of Work in the Twentieth Century* (New York: Monthly Review Press, 1968), esp. chap. 20.

5. See the survey by Kenneth Spenner, "Deciphering Prometheus," *American Sociological Review* 48 (December 1983): 824–37.

6. Note that skill requirements may be increasing, but at a slower rate than skill capabilities or aspirations. The problems of de-skilling and underutilization are separable, as are those of de-skilling and unemployment.

7. See William J. Abernathy, Kim B. Clark, and Alan M. Kantrow, *Industrial Renaissance* (New York: Basic Books, 1983); Joseph A. Limprecht and Robert H. Hayes, "Germany's World-class Manufacturers," *Harvard Business Review* 60, no. 6 (November–December 1982): 137–45; and William J. Abernathy and Kim B. Clark, "Notes on a Trip to Japan," Harvard Business School Working Paper 82–58.

8. In his "Present Tense Technology," parts 1–3, *Democracy* 3, nos. 2–4 (1983).

9. Paul Adler, "Automatisation et travail: le cas des banques" (Ph.D. dissertation, Université de Picardie, 1981).

10. Four French banks figure in Dun and Bradstreet's top ten world banks (by balance sheet).

11. James O'Brien, *The Impact of Computers on Banking* (Boston: Bankers Publishing Co., 1968), cites turnover among bookkeepers in the 40 percent to 80 percent range.

12. See Bright, *Automation and Management*, chap. 5.

13. Bright, "Does Automation Raise Skill Requirements?" 88.

14. See Bright, *Automation and Management*, chap. 4.

15. R. J. Matteis, "The New Back Office Focuses on Customer Service," *Harvard Business Review* 57, no. 2 (March–April 1979): 146–59, recounts the evolution of another product, the international letter of credit: "Where it once took days, 30-odd separate processing steps, 14 people, and a variety of forms, tickets and file holders to process a single letter of credit, it now requires one individual less than a day to receive, issue, and mail out a letter of credit—all via a terminal that is fully on-line to a minicomputer-based system."

16. Each procedure listed in the figure of course involves several suboperations. In bookkeeping alone, some twenty manual procedures are supplanted by computerization. The chart has therefore been simplified so as (1) to keep it within manageable proportions and (2) to present the underlying logic of the sequence of steps common to very different stages in the evolution of the process.

17. Access to the data system is gained by keying in a personal identification code, which also indicates to the computer the nature of the operations the operator is authorized to conduct. It is the use of such codes that makes individualized job monitoring technically possible. In France the threat of labor strife encouraged bank managements to abandon any such monitoring.

18. In reality, some accounts are changed so often that continuous updating of the permanent file is too expensive. These data are fed automatically to locally processed, provisional files, which are then automatically fed into the central, permanent files at the end of the day.

19. Bright, *Automation and Management*, esp. chap. 11.

20. I am relying here on fieldwork conducted principally in one of France's "big four" banks in 1980 and 1981. This research involved several hundred interviews of bank personnel and management at all levels and participant observation for two months as a teller of two branches, representative of stage 3 and stage 4 configurations. See also Adler, "Automatisation et travail," which includes a discussion of the considerable bibliography that has emerged on these themes.

21. Bright, "Does Automation Raise Skill Requirements?" 90.

22. In fairness to Bright, it should be recalled that he was intervening in a context dominated by the fear of massive short-term increases in skilled-labor requirements. He was overstating his case on this subject when he ventured into the broader, longer-run issues we are exploring.

23. B. A. Shiel, "Coping with Complexity" (Palo Alto: Xerox Palo Alto Research Center, 1981). See also Lucy Suchmann and Elanor Wynn, "Procedures and Problems in the Office," *Office: Technology and People* 2 (1984): 133–54.

24. See J. Durraffourg, F. Guérin, F. Jankovsky, and B. Pavard, "Analyse des activités de saisie: correction de données dans l'industrie de la presse," *Travail humain* 42, no. 2 (1979).

25. Bright, *Automation and Management*, 188.

26. See R. Gallé and F. Vatin, *Le modèle de fluidité: étude économique et sociale d'une raffinerie de pétrole* (Bandole: Laboratoire de Conjoncture et de Prospective, 1980); and Elwyn Edwards and Frank Lees, eds., *The Human Operator in Process Control* (London: Taylor & Francis, 1974).

27. On this last factor, see Larry Hirschhorn, "The Post-industrial Labor Process," *New Political Science* 23 (Fall 1981): 11–33.

28. M. O. Ostberg, *Les terminaux d'ordinateur à écran cathodique: problèmes de santé chez les opérateurs* (Paris: Institut National de Recherche et de Securité, Note 1049–86–77, 1977).

29. Bright, "Does Automation Raise Skill Requirements?" 89–90.

30. Computerization, whether of banking or in the form of "flexible manufacturing systems," belies the correlation of automation and rigidity that has been argued by a long tradition running from Joan Woodward, *Industrial Organization: Theory and Practice* (London: Oxford University Press, 1965), to modern production texts, such as Roger Schmenner, *Production/Operations Management* (Chicago: Science Research Associates, 1981).

31. E. B. Fiske, "Computers Alter Lives of Pupils and Teachers," *New York Times*, 4 April 1982.

32. See, for example, Matteis, "The New Back Office."

33. The National Institute of Occupational Safety and Health has produced various publications on these issues.

34. See E. Grandjean and E. Vigliani, eds., *Ergonomic Aspects of Visual Display Terminals: Proceedings of the International Workshop, Milan, March 1980* (London: Taylor & Francis, 1980).

35. Two considerations support this hypothesis: (1) the stand-alone, single-work-station machine is, with automation, replaced by an automatic, integrated system; and (2) the sequential dependence of tasks characteristic of mechanization and exemplified by the assembly line is supplanted by the generalized interdependence of tasks around (interface) and above (design and surveillance) the automatic system. Even though automation permits a certain loosening of the constraints of sequential dependence, especially with modular design principles, the increase in span seems to impel a qualitative shift towards higher levels of interdependence.

36. See Shoshanah Zuboff, "New Worlds of Computer Mediated Work," *Harvard Business Review* 60, no. 5 (September–October 1982): 142–52.

37. This shift generates new legitimacy problems. The criteria for promotion to supervisor level would then give less weight to experience and technical expertise, and more weight to qualities of personality. More subjective criteria for promotion would mean more fragile authority.

5 University and Industry Cooperation in Microelectronics Research

F. Dana Robinson

As the preface to this volume makes clear, state policy-makers have targeted university research and training programs as vehicles for assisting local microelectronics firms and attracting new ones. Such policies are being implemented on a substantial scale in California, Arizona, Texas, North Carolina, New York, and Minnesota. Generally, microelectronics firms have played leading roles in campaigns for increased state support of university programs and have contributed generously themselves. It is reasonable, therefore, for economic development planners to perceive links between the quality of local university research and training efforts and the prospects of the local microelectronics industry.

This chapter analyzes the nature and importance of these links. To some extent, the issues are straightforward. For example, the microelectronics industry is heavily dependent upon the skills of its scientists and engineers, and recruits competitively for the top talent. The proximity of first-class universities simplifies the recruitment of graduating students and gives employees opportunities to continue their education and keep up with developments in their fields.

Many firms also have expectations that efforts to improve the scale, scope, and quality of university research will pay off in the form of discoveries that facilitate innovation. Because they are willing to underwrite part of the cost of such efforts, this expectation must be taken seriously, especially by economic development planners who seek ways to strengthen local microelectronics producers. However, the issues here are less straightforward than in the case of human capital investments.

Interactions between firms and university researchers in which firms hire faculty members as consultants or sponsor specific research projects at universities have long been characteristic of the microelectronics industry. In such cases the relationship is bilateral, and sponsored research is conducted according to relatively explicit specifications concerning scope, responsibilities of researchers, compensation, and property rights. In contrast, in today's ambitious efforts to upgrade universities, firms participate multilaterally and without definite commitments from faculty members concerning their future research activity. Expectations of benefits are therefore much more open-ended than in the past. The microelectronics industry is now willing to contribute to cooperative undertakings that benefit the entire industry, or regional segments of it, rather than single firms alone.

The willingness of American microelectronics firms to cooperate in strengthening university research and training programs coincides with their alarm at the emergence of Japanese firms as formidable competitors.[1] To analyze this new concern with university research without taking into account anxiety about overseas competition would be to miss the point. Were it not for Japan's highly credible bid for competitive parity with American firms in world microelectronics markets, it is unlikely that dissatisfaction with the current state of basic research would motivate firms to mount a major effort to subsidize universities. After all, the system that has evolved over the past thirty years to support research and training has contributed to an American preeminence in the industry that continues to this day.

Leading American universities have housed dynamic research and training programs in fields related to microelectronics since the 1950s. The Department of Defense, the National Science Foundation, the National Bureau of Standards, and to a lesser extent the National Aeronautics and Space Administration also aid the research and training effort.[2] Before its deregulation AT&T allocated, with regulatory approval, a portion of its revenues to extensive programs of basic research in microelectronics at Bell Laboratories.

The microelectronics industry is not merely a passive consumer of the scientific information produced by universities. Research, innovation, and enhancement of the expertise of researchers occur at high rates within firms themselves. The semiconductor industry, which manufactures microelectronic devices for sale to the open market, devotes about 9 percent of sales revenues to research and development, compared with an average of less than 3 percent for all U.S. manufacturing firms.[3] Microelectronics firms are clearly committed to innovation as a corporate goal and are willing to invest heavily in research and product development strategies.

Although industrial research and development is predominantly oriented to the use of existing scientific and technical expertise in developing commercial products, IBM, AT&T, and other leading firms also conduct extensive programs in basic scientific research. Because firms look to their internal research units for the knowledge needed to implement corporate strategies based upon technological innovation, it remains to be seen how the expansion of university-based research could expand firms' ability to innovate.

By challenging American expectations of continuing global hegemony in microelectronics, competition from Japan has undermined complacency concerning basic research and training. The absolute volume of U.S. expenditures in these areas, already quite large, is increasingly viewed as inadequate. More interesting from the standpoint of public policy is a new perception that current research and training is not structured to contribute as much to industrial performance as would be possible under alternative organizational arrangements. In particular, there is a growing recognition that the rarity of joint research and development projects in the industry restricts the rate of expansion of the common knowledge base.

Until recently, the entrepreneurial, intensely competitive nature of the microelectronics industry was viewed as a key factor in its technological leadership. The emergence of Japanese firms as major competitors in markets for state-of-the-art products has, however, now caused analysts within and outside the industry to focus anxiously upon the apparent cooperation, coordination, and support characterizing relationships between Japanese firms and their government. This overseas competition has certainly made American firms and universities more receptive to novel ways of organizing research in microelectronics. One result has been the recent foundation of cooperatively supported university-based microelectronics research centers, as well as the Microelectronics and Computer Technology Corporation and the Very High-speed Integrated Circuit Program sponsored by the Department of Defense. William Norris, chairman of Control Data Corporation, is a conspicuous champion of joint research and development, and has been joined by other industry leaders.[4]

Traditionally, academic research has taken place without close coordination with the industry's research and development efforts. To be useful, university research must somehow complement corporate research and development without the coordination afforded by firms' internal organization. Academic research is not likely to serve as a substitute for research conducted by industrial scientists and engineers who actually develop innovations. Rather, the importance of university research will lie in specialized areas that are only loosely coupled to firms' efforts to define and develop proprietary technologies.[5] This chapter identifies several areas where university research is likely to contribute to innovations in microelectronics.

Miniaturization and Microelectronics Innovation

The U.S. microelectronics industry came into being in the early 1950s, when the first transistors and diodes appeared on the market. Transistors and diodes superseded the vacuum tubes that once performed amplifying and digital switching functions in radios and the earliest computers. Unlike the vacuum tubes, they are solid-state components made of semiconductors: silicon, germanium, and, in a few instances recently, more exotic materials. Semiconductors have the advantage of consuming far less power, producing less heat, operating more reliably, and, above all, permitting devices to be made ever smaller.

Miniaturization has been an important attribute of semiconductor components ever since transistors were developed for use in hearing aids, which were among the first commercial applications of semiconductors. With the planar process, miniaturization became a central element of manufacturing strategies as well. The planar process, commercialized in 1959, facilitated low-cost batch processing by allowing the simultaneous formation of several dozen transistors and diodes upon wafers of silicon. The technique is complex, but can be likened to a silkscreen process: selected materials are deposited onto the silicon through a template. Successive use of appropriate templates to apply different materials builds up components that can eventually be separated when processing is complete.

Degree of miniaturization is highly correlated with important product attributes such as unit processing costs, operating speed, system operating reliability, and system requirements for space and cooling. With few exceptions (as in military equipment) the greater the density of integrated circuits, the more advanced—innovative—they are. Similarly, process techniques are innovative insofar as they can be used to produce smaller working elements upon chips. This relationship between innovation and miniaturization has been constant in the history of the industry.

> As the variety of applications grew, and as integrated circuits very gradually replaced discrete devices, the course of further technical development was quite clear. There was none of the profound interaction between scientific theory and technological practice that had characterized the first years of the development of the transistor. The integrated circuit concept did not rest on any novel application of scientific theory; it was an engineering achievement. Moreover, the fundamental process technology was in place. Hence, a simple engineering heuristic dominated the course of technology over the next two decades: make it smaller.[6]

Even so, it would certainly be incorrect to conclude that there was anything easy about transforming the transistors and diodes of the 1950s into microcomponents so small that now several hundred thousand can be packed onto a single silicon chip. In fact, microfabrication has been described as perhaps the most difficult and demanding of all industrial processes.[7] "The secret of success in semiconductors is to be able to make literally millions of devices without a high proportion of the circuits being defective. But the causes of defects are not always understood nor are they easy to control when they are understood."[8] Not only must each step in the microfabrication process occur within very demanding specifications, but the overall number of steps is large and the appearance of defects at a given stage may reveal problems with roots in earlier stages. The result, according to one scientist, is that "seemingly identical apparatuses and starting materials often yield different results. Recipes must be carefully followed, with little understanding of which aspects of a process are critical or what contaminants are important, and why."[9] Not surprisingly, leading microelectronics producers conduct extensive internal research and development to enhance their efforts to successfully produce defect-free innovations.[10]

The rapid rate of progress of miniaturization in a technology involving well over one hundred steps and highly sophisticated equipment indicates not ease of innovation, but impressive skills on the part of firms in directing and coordinating a process in which large teams of specialists from a variety of backgrounds achieve major improvements in the output of a highly complex technology. In sum, successful innovations result less from scientific discoveries than from close communication between the researchers investigating new technologies and the managers who will ultimately have to translate research and development into reliable, high-volume manufacturing operations. The expertise that supports a steady expansion of capabilities in miniaturization is rooted in extensive experience with a complex manufacturing process, and the ability to manage changes in it.

University researchers lack technical and managerial experience with manufacturing technology. Moreover, their own laboratory facilities tend to lack the costly, sophisticated equipment with which state-of-the-art integrated circuits are produced. It is thus unlikely that university researchers will invent advanced integrated circuits that could strengthen the product lines of U.S. microelectronics firms. They are not equipped to deploy the technology needed to design prototypes of integrated circuits, and the processes used to make them in university laboratories are not those that firms use. What, then, do firms expect to learn from the university research programs they are supporting?

Although integrated circuit manufacturing itself requires elaborate industrial processes, many of these processes depend upon specialized equipment and techniques that can be developed with a minimum of feedback from manufacturing operations. For simplicity, these can be termed *production inputs*. Such inputs include hardware, software (principally for computer-aided design of chips), and scientific concepts not embodied in technology. Microelectronics firms themselves conduct extensive research and development in these areas. In-house researchers have the advantage of regular contacts with manufacturing operations and the opportunity to test experimental versions of process equipment and other techniques in a production setting. Under what circumstances are firms better served by funding university researchers than by conducting their own investigations of inputs?

Firms' in-house research and development units are in a strong position, relative to universities, to develop input innovations that represent *incremental improvements* in currently used technology. But, conversely, prior experience with their firms' manufacturing technology would be less useful in the development of *radical departures* from existing input types. Industrial researchers' ability to communicate with input users in manufacturing would contribute less to their investigations, because the shared technical assumptions underlying internal communication would not necessarily be relevant to the investigation of major innovations.

In fact, a focus on incremental technological improvements rather than more far-reaching experimental research and development is evident in a number of inputs that semiconductor firms develop internally. The all-important handling of silicon purification and crystal formation, for example, is still based largely on empirical experience rather than scientific theory and methodologies.[11] Similarly, though efforts by integrated circuit producers and equipment suppliers to develop new microfabrication machinery have produced a number of highly sophisticated approaches, outside observers still remark that the technology development policies of the merchant semiconductor industry appear to emphasize incremental changes in current equipment more than is warranted by the challenging agenda of research issues connected with further miniaturization.[12]

The point is that internal research and development taking advantage of extensive feedback from manufacturing operations is likely to lend itself well to the development of incremental improvements in inputs, and therefore be used for this purpose, because of superior communication and coordination. These advantages are less relevant in the case of experimental research, in which a broader range of scientific and technological possibilities is considered. In fact, constant intraorganizational contact with manufacturing personnel can produce a bias toward possibilities already associated with the firms' particular operations; this can compromise the effectiveness of experimental research and development even when firms recognize its importance.[13]

This tendency to stick with familiar techniques is a serious management issue for firms in a highly competitive, innovative industry. Strategies are needed to identify areas where radical innovation is likely and to plan responses. University research can aid managers in choosing between different approaches to innovation, as well as providing data that aid in actually developing new versions of inputs.

The expectation that expansions in university research will improve firms' ability to innovate thus implies a recognition that the microelectronics innovations of this decade will not be based entirely on evolutionary progress in existing inputs. Discussion in the following pages identifies areas in the development and production of integrated circuits where alternative approaches to innovation may prove worthwhile. Academic research can contribute to such innovations in several ways. Four in particular should be kept in mind.

First, some innovations in inputs will result from the expansion of knowledge bases that are quite separate from expertise in miniaturization per se. For example, improvements in computer-aided design (CAD) of complex integrated circuitry will require significant progress in computer science. This does not mean that microelectronics firms do not also have expertise in such areas; with integrated circuit design, they obviously do. But when additions to a largely independent knowledge base produce innovations, the manufacturing expertise of microelectronics firms contributes correspondingly less to the success of the process. Alternative approaches developed by outside specialists are more likely to offer compelling new solutions.

Second, some innovations in input require relatively close interactions with basic scientific research. Progress in transistor technology in the 1940s and 1950s, before the planar process established a more predictable technological regime, is a case in point. Innovations in integrated circuit packaging, which are extremely important in systems development, do not depend on advances in basic science. But improvements in certain other inputs run into obstacles caused by a lack of scientific understanding of underlying phenomena. Under such circumstances an orientation to academic research rather than technological applications may be valuable.

Third, experimental development of some inputs involves multiple approaches whose comparative advantages are presently uncertain. In such contexts confidence in the continuing enhancement of current procedures is unwarranted, because a quite different approach may prove to have major advantages.

Fourth, certain areas of microelectronics technology present bottlenecks where further progress in miniaturization is unusually difficult. The design of complex, high-performance circuits is a good case in point: CAD is as yet unable to reduce the tremendous demands on skilled manpower created there. Bottlenecks of this sort suggest that attempts to achieve further miniaturization with existing techniques may yield diminishing returns.[14]

Input Innovations in Microelectronics: An Overview

Innovation in inputs can occur separately from the evolution of miniaturization capabilities in the manufacturing process. Table 1 shows a breakdown of inputs in the development and production of integrated circuits and groups them according to three general phases, of which manufacture is the last. The overview presented in this section is designed to convey the potential for innovation at each point in the process. Further discussion considers the possibilities in light of the four ways, identified above, in which academic research can especially aid in innovation.

Today's integrated circuits are miniaturized electronic systems, and many innovative inputs are directed at systems design. Systems design involves both the configuration of the individual microcomponents on the integrated circuit and the circuit design that links microcomponents into an electronic system. These activities are an aspect of miniaturization separate from the technical capability to make microcomponents smaller. That is, by designing new transistors requiring less surface area and by increasing the functional capabilities of circuits, innovations in integrated circuitry can be achieved without awaiting further innovations in the manufacturing process.[15]

Innovations in systems design often rest on advanced knowledge of mathematics and scientific theory. Microcomponents include individual transistors, diodes, resistors, and the tiny wires that provide electrical connections between them. As miniaturization has progressed and the components have been shrunk to structures whose widths are on the order of one micron (one thousandth of a millimeter), their electrical behavior has become less and less a function of bulk properties of semiconducting materials.[16] Instead our current understanding of such behavior begins with mathematical models based on advances in solid state physics in the theory of electrical conductivity in highly miniaturized structures. Modeling and simulation also play essential roles in circuit design. Complex integrated circuits with more than 100,000 transistors simply cannot be designed by hand; CAD software has evolved over the years to automate relatively routine functional design, layout, debugging, and performance simulation. In short, the design of both transistors (at the micro level) and circuits (at the macro level) involves innovations as distinct inputs with their own knowledge bases and technologies.

Logically the next inputs to be considered are those concerned with the interface between systems design and the physical fabrication process. This interface involves a sophisticated set of technologies that can be loosely grouped under the term "pattern transfer." The first component consists of masks—the "templates" alluded to above. Integrated circuits are formed in a sequential process in which microcomponents and their interconnections are built up, layer by layer, upon the surface of silicon wafers. Each layer is defined by a separate mask, which is a miniaturized version of the initial blueprint used to define the structure of the integrated circuit. The mask transfers patterns by controlling the illumination of selected areas of the wafer's surface, which is periodically coated with photosensitive chemicals. Light exposes "windows" in this surface coating, through which the underlying portion of the chip can be modified chemically. Additives (dopants) can thus be diffused to control the electrical functions of the silicon; metal films are put in place to form interconnections; the silicon can be selectively oxidized to create insulation.

Innovations at this stage center on increased precision in pattern transfer. For example, at feature widths of one micron, resolution becomes a problem, in terms of both the clarity of the pattern illuminated on the wafer and the responsiveness of the photosensitive chemicals. Because the wave length of visible light yields insufficient clarity, the development of equipment that projects other forms of radiation (such as ultraviolet or x-ray) likewise constitutes an aspect of innovation in pattern transfer.

Table 1 groups systems design and pattern transfer technologies as "inputs developed before microstructure formation," because they must be in place (in either experimental or commercialized form) before research and development concerning further miniaturization can occur. Improvements in the size, structural regularity, and chemical purity of single-crystal silicon wafers are similarly categorized.

"Inputs to microstructure formation," the second of the input groups shown in table 1, concern the process of learning how new types of

Table 1. Microelectronics Processing Innovations: Inputs Requiring Modification

Input	Stage
Design Aids and Models	INPUTS DEVELOPED BEFORE MICROSTRUCTURE FORMATION
new ultraminiaturized transistor configurations	
computer-aided design for layouts of complex ICs	
Pattern Transfer	
projectors to transfer patterns to chips, chemical resists	
Materials	
purer silicon	
larger silicon wafers on which ICs are created	
epitaxy, etching, diffusion, and oxidation to create individual IC components	INPUTS TO MICROSTRUCTURE FORMATION
new gate electrode and interconnection materials	
Testing and Microanalysis	
assessment of structure and chemistry of microcomponents	
computer programs to check complex ICs for defects	INPUTS TO IC MANUFACTURING
Packaging of Chips	
Factory Automation	
for greater control over fabrication process	

transistors and other microcomponents can actually be built. At issue here are such attributes as the speed at which new microcomponents conduct and switch electrical signals, the reliability with which they function, and the likelihood of defects arising during processing. Materials analysis and modeling of the physical processes underlying the behavior of given microstructures are key techniques of research and development at this stage. Closely linked with materials analysis is the development of the technologies used to transform the exposed surface of the silicon wafer. These include chemical materials, chiefly etching substances that penetrate the top surface of the wafer through windows in the resist, and the thin films of materials from which microstructures and their interconnections are built up, but also the various types of equipment for diffusing dopants into the silicon.

The development of new microstructures is an essential part of miniaturization that falls between two well-defined industrial activities, that of producing inputs and that of integrated circuit manufacturing itself. It falls outside the first group because new microstructures are not marketable inputs but prototypical constituents of integrated circuits. On the other hand, their development is largely a modeling activity that does not require the deployment of a full-blown integrated circuit production line. But at this stage links to manufacturing and marketing are obviously needed to define critical performance dimensions and the range within which attributes must score. For this reason such research and development generally takes place within the firm. Nevertheless university research has also made important contributions in this area.[17]

The third phase, "inputs to manufacturing," concerns the volume production of integrated circuits. Difficulties in efficient production of defect-free, highly miniaturized integrated circuits are increasingly creating a demand for a systems-oriented approach to computer-aided manufacturing (CAM) that is separate from the manufacturing technologies currently in place. Firms (such as General Electric) that produce industrial process controls may become important sources of factory automation systems.[18] Moreover, organizers of microelectronics research centers at Stanford and MIT and in North Carolina have set a high priority on securing advanced integrated circuit fabrication equipment comparable to that used in industry.[19] This may stimulate academic interest in CAM and the creation of software and monitoring devices that will be relevant to industrial needs.

This overview suggests that although microelectronics firms are investigating the development of technical inputs, such research and development is often quite removed from the development of proprietary expertise in integrated circuit production. University research may be a useful alternative to internal investigations of the inputs. As suggested earlier, its comparative usefulness is greatest in resolving technological bottlenecks, exploring unanswered theoretical issues concerning underlying natural phenomena, and evaluating multiple approaches to innovations. The groups of inputs shown in table 1 are next examined from these perspectives.

Inputs Developed before Microstructure Formation

The first group of technologies, inputs developed before microstructure formation, consists primarily of inputs to the design of circuits and to their transfer to the surface of the silicon wafer. The patterning process begins with conceptual micro and macro models of circuit behavior. As the pattern transfer process comes to involve actual transformation of the wafer surface, inputs become increasingly embodied in physical technologies. Bottlenecks have induced technological innovation at many stages of this process, but these adjustments have not always been optimal. For design modeling, the problem is a lack of basic scientific theory. For pattern transfer technologies, there has been no lack of new approaches, but further experimentation is needed to establish their comparative value.

Bottlenecks in circuit design have been a recurrent problem since the days of the transistor, when the complexities of wiring computer backboards were a major impediment to computer innovation.[20] Miniaturization transferred the problem to integrated circuit production. Computer-aided design (CAD) has evolved in piecemeal fashion to help designers manage the complexity of connecting tens of thousands of microcomponents. Often using systems developed by specialized suppliers, firms have been able partially to automate such procedures as the definition of the logical operations needed to make the chips function, the layout of circuitry for implementing logic, the debugging of circuit layouts, the transfer of layouts to masks, and tests for defects. However, these systems are only partial and incomplete aids to design. Researchers are now attempting to develop a rigorous approach to the theory of designing complex systems so that the software used to define basic functional characteristics of integrated circuits will also automatically debug circuits and layouts and control the physical representation of the design.[21]

Even without such an integrated approach, microelectronics firms are performing prodigious feats in the design of complex integrated circuits. The microprocessor chip that made it possible for Hewlett-Packard to introduce a desktop unit with the power of a small main-frame computer has 450,000 transistors and other microcomponents on an area the size of a thumbtack.[22] Another innovation widely regarded as a design triumph has been Intel's "microminiaturized main-frame," which puts the equivalent of a state-of-the-art mainframe computer's central processing unit onto three chips. The major reduction in hardware costs and space requirements holds out great potential for extensive programmed control of robots and other industrial equipment.[23] These advances in microprocessor technology are expected to have major effects on office automation (by speeding the development of work stations) and factory automation.

The historical pattern in microelectronics has been that exceptionally proficient firms, like Intel, advance the industry's technologies by overcoming extremely difficult technical problems and dominating the resulting supply of advanced integrated circuits. This pattern may characterize design innovations, but it cannot be regarded as an optimal solution for the microelectronics industry as a whole. As microelectronics devices have become more pervasive, the number of firms with bases in other industries that have become involved in developing and manufacturing integrated circuits has necessarily grown. For example, many firms developing office work stations and industrial robots may see benefits in designing their own microprocessors, yet lack the unusual expertise of firms like Intel. Their relatively inefficient, costly efforts to create systems around microprocessors would benefit from a more systematic approach to integrated circuit design.[24]

Given the lack of such a systematic approach it is not surprising that innovators are experimenting with a multitude of alternative techniques, involving both concepts in computer science and choices in product attributes of integrated circuits. For example, a silicon compiler program developed at the California Institute of Technology is aimed at minimizing design and transfer problems related to the complex transistor interconnections on integrated circuits. This is accomplished through a reliance on integrated circuits with a relatively standardized, repetitive configuration of circuit elements. The customized design features chosen by the compiler can be incorporated during the last stages of processing.[25] Other approaches are also possible.[26] Rather than designing around highly regular circuit layouts, a relatively small set of functional "building blocks" can be defined. Circuit design would thus be simplified to the problem of defining interconnections between the blocks. Artificial intelligence concepts, the MacPitts LISP compiler, and an approach based on structured software design are also being developed. In each case, separate compilers, interface definitions, interactive graphics, transistor models, simulation programs, timing checks, verification techniques, and data base systems are involved. Evaluating the comparative advantages of competing design approaches is thus extremely difficult.

Such design problems at the macro level receive extensive attention because their effects are felt directly, in burdensome manpower requirements and difficulties in optimizing the performance of integrated circuits and the larger systems that depend upon them. But at the micro level, the longer-term horizons of basic research obscure the significance of bottlenecks—and these are likely to become increasingly important during the last half of the decade, when progress in miniaturization will involve microstructures with widths of less than a micron.

Micro level design of microcomponents requires a theoretical knowledge base for predicting electrical performance on the basis of various parameters. Heretofore, performance models found to be valid at a given level of miniaturization could be adjusted, or "scaled," to simulate performance at a greater level of miniaturization. Scaling in models assumes that underlying physical processes themselves are not changed by miniaturization. It is widely believed that submicron-width microcomponents invalidate this assumption, because "properties of microminiaturized structures are not generally deducible from the bulk properties of their composite constituents."[27] For these tiny structures, deviations from "bulk properties" become important determinants of performance. At so small a scale, electrical characteristics on the surfaces of microcomponents come to influence current flows, and interfaces between the doped silicon, interconnections, and insulators also contribute to performance in new ways. Defects and other problems in reliability become paramount concerns. The bottleneck foreseen by the National Research Council in its report on basic research needs in microelectronics is summarized in its conclusion that "our basic understanding of scaling-associated changes in the properties of materials is primitive."[28]

A final area of uncertainty in basic research involves the creation of single-crystal silicon wafers whose purity and physical structure are suitable for submicron microcomponent widths. At present, the "growth" of such crystals is understood on the basis of extensive empirical experience rather than scientific theory.[29] This creates a bottleneck to further innovation for two reasons. First, miniaturization is not the only route to greater performance. Circuit density can be increased simply by making the integrated circuit bigger and adding circuits.

Because unit costs are proportional to the number of integrated circuits formed on each wafer, larger integrated circuits mean larger unit costs, unless wafer diameter itself can be increased. Wafer diameters have increased in four stages since 1965.[30] The limiting factor has been the ability to control impurities and irregularities in the structure of the crystal.[31] Second, miniaturization of microcomponents has decreased the acceptable ranges for impurities and structural discontinuities. For example, future integrated circuits will require wafers with at most a 1 percent variation in electrical resistivity from point to point; currently the best commercially available silicon has a variation of 5 percent to 10 percent.[32]

As for the projection equipment that selectively illuminates the wafer surface through the masks, current knowledge has thus far been equal to the task of developing responses to emergent bottlenecks. The problem here, rather, lies in evaluating the comparative advantages of new types of equipment. The photolithographic process for exposing chemical resists has employed visible light since the introduction of the planar process. Visible light can expose areas as narrow as one micron in width, a limit that commercial technology is now reaching; narrow patterns become unacceptably fuzzy due to the diffraction of light. The initial response to this problem has been to modify optical projection equipment to operate in the ultraviolet spectrum, where diffraction occurs at a much smaller width (0.3 micron).[33] Eventually the quest for further miniaturization will lead to adoption of radiation that is even less affected by diffraction, such as electron or ion beams and x-rays. These technologies are in different stages of development, with major bottlenecks still to be overcome. For example, electron-beam direct-write projectors are as yet too slow to be used for mass production of integrated circuits.

Adequate comparisons of innovative pattern transfer equipment are likely to be complicated. The technologies associated with the various beams are quite different, and each requires a special resist sensitive to its radiation. Moreover, equipment can operate in one or more modes (direct-write, projection, proximity, or contact). It is also used for different tasks: to make masks, and to illuminate wafers at different stages of integrated circuit processing.

Electron-beam projectors are becoming commercially available and, to many, promise important contributions to the miniaturization process. In the opinion of at least one consultant, however, late in 1982, "Electron beams died two years ago, but the smell of the corpse is only now reaching the board rooms. The hopes for writing with electron beams [are] a classic case of becoming enamored with a technology whose time just isn't coming."[34] The lesson from this evident lack of consensus is that it would be risky for firms to rely solely on internal experience with a given approach for information concerning its long-term usefulness.

Inputs to Microstructure Formation

Researchers who are attempting to create prototypes of highly miniaturized microstructures are hampered by bottlenecks in materials science.

> It is quite simple for the physicist or engineer to argue that current integrated circuits can be scaled down to the few tenths of a micron level, but it is quite another matter to accomplish this scaling because it requires improvements in the quality of substrates, deposited films, resists and implants. Consequently, materials science in many ways sets the pace for microfabrication.[35]

Two principal sources of uncertainty confront firms that must maintain an expertise in materials science.

First, testing laboratory prototypes depends upon precise analysis (characterization) of the microstructures that have been formed. The techniques and equipment used for materials characterization in the past are not adequate for submicron structures.

> The problem is magnified because the structures are physically small, the properties are often configuration-dependent and frequently are controlled by shielded interfaces, and finally our interest is often focused on nonequilibrium materials systems (composites) rather than on a single material. As tools we have electrical techniques that make use of the device as a probe. . . . However, these often identify a problem but do not define the cause or indicate a cure.
>
> Here we must rely on old and new analytical techniques adapted for use at microdimensions. These techniques involve elucidation of chemical, structural and topographic aspects of the microstructure in question and are impeded by noise problems when the probe used becomes very small. Consequently, we must evolve better sensing techniques and system component matching to enhance signal levels while not increasing noise levels. In medicine, where microanalysis is becoming more pervasive, considerable progress is being made in nondestructive three-di-

> mensional analysis on living organisms, where both of these requirements are paramount. One might cite the use of advanced imaging systems or of computer-aided tomography (CAT) for the real-time imaging of buried structures. Such procedures should be developed for this field, permitting three-dimensional study of materials systems without destructive interference.[36]

Second, the processing of silicon wafers that creates microstructures includes techniques that cannot be adapted to submicron dimensions. Bottlenecks chiefly occur at the stage when the surface of the silicon is etched through the exposed areas of the chemical resist and when chemical films are deposited upon the surface. Etching has long relied upon a wet chemical process in which the wafer is bathed in a corrosive solution. This technique cannot define features with the exactitude needed for ultraminiaturized microcomponents, and other methods must be developed.[37] Similarly, the high-temperature processes long used to deposit thin films upon wafers involve now unacceptable degradations of surface regularity and reductions in clarity of features. In response, new techniques and new materials, particularly for the creation of thin films upon microstructures and interconnections, are under investigation. The comparative advantages of these different approaches have not yet been completely determined.[38]

Inputs to Integrated Circuit Manufacturing

The new types of pattern transfer and wafer fabrication equipment needed for microfabrication are much more sophisticated and costly than earlier versions. Moreover, the worldwide open market and captive demand for integrated circuits is projected to increase rapidly.[39] Thus firms committed to maintaining market share must continually expand capacity. The resulting strain on the capital resources of American firms has received widespread attention. Concern over these burdensome capital requirements has focused on the fact that integrated circuit manufacturing in the United States is not highly automated. According to one survey, "the investment committed by U.S. firms and the U.S. government to the semiconductor factory of the future is small. The degree of automation in the U.S. today is limited to the IBM QTAT line, with other companies approaching the concept from various angles, with focus on pieces rather than the whole."[40]

Automation holds the promise of improving capital productivity through better inventory control and materials flow. Academic research is likely to contribute to progress in computer-aided manufacturing (CAM) to improve the yield of defect-free integrated circuits.[41] Defects can arise for many reasons. Dust and other contaminants can easily ruin integrated circuits, as can variations in humidity and temperature, vibrations, and other perturbations that influence the chemical and structural nature of microstructures. CAM begins with materials science research to reduce the present uncertainty concerning microstructure formation. Proper characterization of the processes employed in integrated fabrication permits a fuller understanding of the causes of defects in microstructures. Automation of the manufacturing process allows automatic adjustments in processing operations that will maintain undesirable variables at acceptable levels.

To create the automated testing and sensing equipment that monitors factory operations and returns feedback to process controllers, extensive research and development is required involving the entire spectrum of microelectronics expertise. For example, development of monitoring equipment and sensors would require expertise in materials science. Creating the complex software for translating feedback from processing to adjustments in operating parameters would require expertise in the manufacturing phase. However, such software would also need to be linked to the CAD systems used to design circuits.[42] And a great deal of specialized effort would be required to develop equipment that could be fine-tuned to a high degree of precision.

Universities' Responsiveness to Industry Research Needs

The selection and evaluation of new inputs involves a number of activities. A useful distinction here is that between *horizontal* aspects of the conduct of research and development, which involve interactions with the industry's knowledge base as a whole, and *vertical* aspects, which involve interactions with users of research information for product development, manufacturing, and marketing.[43]

Horizontal aspects of research include, first, the selection of a promising research area from among the possible approaches to developing a particular type of input; if, after consideration, none appears promising, a new alternative may have to be found. Second, after a particular input has been chosen for study, relevant work of other researchers needs to be examined. Third, in specifying a research design, ancillary requirements such as measurement procedures, equipment, software, and materials need to be assessed. Fourth, data and conclusions from experiments need to be compared with similar research elsewhere.

Probably the most important vertical aspect of research is determining the performance variables that are important to users. For example, development of an approach to CAD that sought to minimize the complexity of integrated circuitry but allowed highly wasteful use of a chip's surface area would be of only academic interest until some attention was paid to reducing space requirements. Researchers must also ascertain thresholds in performance that can meet users' requirements. Third, results must be communicated to potential users, and feedback from them must be made available to researchers. Fourth, ancillary requirements must be met. Researchers must be supplied with specialized, and often costly, research and production equipment, as well as relevant data bases and software. Fifth, because specialized research perspectives are inadequate to cope with many problems in microelectronics, researchers need access to colleagues working in related fields. The establishment, on a formal or informal basis, of such teams of specialists implies that an organization has some kind of vertical orientation to potential users.

These aspects of research into new inputs suggest a set of corresponding criteria, or tests, for judging the efficiency with which particular institutions organize such research and development.

HORIZONTAL

- Are researchers familiar with the leading edge of research concerning particular types of inputs? Do they have access to current information of a comparative nature concerning the most pressing issues associated with the range of possible approaches to input development?
- Will researchers develop needed ancillary aspects of evaluative research such as new measurement equipment and procedures?
- Can research discoveries be shared with peers to solicit feedback concerning quality and general applicability?

VERTICAL

- Are researchers aware of, and motivated to consider, performance features of materials, equipment CAD systems, and models of microstructure behavior that firms consider critical to determining technological feasibility?
- In setting research priorities, are researchers guided by critical values that can diagnose bottlenecks?

- Can research results be communicated effectively to users, and can those users play a role in guiding research activity?
- Are the completion dates of experiments consistent with firms' planning horizons for evaluating new inputs?
- What is the likelihood that a "critical mass" of researchers from related fields will be available to make multidisciplinary research possible?
- Do available equipment, software, computer links, and other facilities permit researchers to respond to research issues raised by state-of-the-art industrial research and development?

Universities and the research and development units of microelectronics firms meet these criteria with varying degrees of success.[44]

Evidently, few existing research organizations enjoy institutional characteristics that allow them to satisfy all of these criteria well. IBM and Bell Laboratories come closest: they have abundant resources and can employ numerous researchers with expertise in all phases of research and development, from pure scientific research to development of technology. Many staff members participate in professional organizations of a horizontal nature, along with academic researchers working in the same disciplines. This suggests that two of the horizontal criteria are adequately met: staff are likely to be aware of relevant work occurring elsewhere, and they have the opportunity to solicit feedback from peers. As for the third, resources of these large organizations do permit the development of ancillary input; Bell Laboratories' development of molecular-beam epitaxy research equipment is a prominent example.[45]

Researchers at Bell and IBM also have extensive opportunities for vertical interactions with users. At IBM such contacts appear to have stimulated creative extensions of both scientific knowledge and technological capabilities.[46] How well captive microelectronics research units meet the input assessment needs of the Bell System and IBM is not easy to analyze, but in one view, "IBM is . . . (in many instances) on the leading edge of technology in both process and product."[47] Certainly both organizations enjoy a critical mass in their teams of microelectronics researchers and have no difficulty in providing them with needed research inputs.

IBM and AT&T are in a class by themselves in the funding of scientific research. Other firms are much more constrained in their ability to support research that looks to professional organizations and literature for ideas, rather than attending strictly to internal needs. The microelectronics industry is so intensely competitive that firms must move rapidly in the development of applied technology: vertically oriented responses to problems in development and manufacturing technology necessarily take precedence over professional activities concerned with research taking place in other organizations.

In individual firms research and development connected with the evaluation of inputs often suffers from inefficiencies because of problems in satisfying horizontal criteria. Researchers may be extremely competent in addressing problems connected with their firms' own technologies but not be in a position to do comparative research over a broader spectrum. Mitigating these inefficiencies has been the usually free exchange of information between firms located in Silicon Valley, where employee mobility is high and geographical proximity allows personal relationships to flourish between employees of different firms. One quip has it that the Wagonwheel Bar there served as the "fountainhead of the semiconductor industry."[48] But many microelectronics firms are not located in Silicon Valley, including the two leading merchant producers and many firms with captive units. And as scientific theory (relative to technological experience) becomes increasingly important for research in highly miniaturized technologies, firms will need to exchange information systematically through professional organizations as well as through informal contacts.

By contrast, universities have the potential to conduct evaluative research with a high degree of efficiency in terms of horizontal criteria. Academic researchers routinely read professional journals and, through participation in the meetings of professional societies, keep up personal contacts with researchers in other organizations. The emphasis on publication in determining academic promotions encourages horizontal flows of information; secrecy about research activity taking place at a particular organization is rare. The professional rewards associated with publication also provide incentives for university-based researchers to develop generic aids to research, such as new materials analysis techniques and instrumentation, or scientific models of microstructure performance, whose value as marketable innovations may be doubtful or nonexistent. Institutions such as peer review for journal publication and the meetings of professional societies provide researchers with feedback concerning the quality of their work and its potential for further discoveries.

How easily university research may be able to satisfy vertical criteria is much less clear. In particular, academicians do not have strong professional incentives to keep informed of industrial research and development and the performance variables that firms consider critical, unless these happen to overlap with research interests considered important by peers in their own field. Their own work will be accepted for publication if favorably reviewed by peers, whether or not the findings are relevant to industry.

Of course, university researchers do not spend their time entirely as autonomous investigators at the frontiers of research in their chosen fields. They also operate collectively as research faculties that must compete with other groups within their universities for funds, and with other university faculties for prestige, government and private funds, and (in a tight labor market) top researchers. In recent years vertical interactions have featured in this competition in various ways. Universities are attempting to accommodate the multidisciplinary nature of many aspects of microelectronics research and to respond to the need for a critical mass of researchers with a broad range of research interests.

Berkeley, the first university to establish a center explicitly devoted to microelectronics research (in 1960), was long unique in possessing a facility that was open to faculty and students from many disciplines.[49] In the past several years other universities have established microelectronics research centers (MRCs), often equipping laboratories with up-to-date testing and processing equipment similar to that used in industry. These developments are necessary if university research is to be responsive to industrial needs. The neglect of vertical interaction in the past is reflected in one commentator's conclusion that "in general, academic research has not kept pace with industrial research and development in contributing to the technology base of integrated circuits."[50]

The establishment of microelectronics research centers at universities, even with support from industry, does not in itself guarantee responsiveness to industrial research needs. Two special characteristics of university research are at issue here. First, experimental research involves unpredictable findings, dead ends, and serendipitous discoveries of new phenomena. MRCs cannot make detailed promises to firms concerning the approaches to be taken to research problems or the areas where important discoveries will occur, because these are unknown at the outset. Second, academic researchers value their autonomy. Renaming a laboratory building the "Microelectronics Research Center" does not guarantee that scientists and engineers from different fields will work together. Moreover, promises by administrators of such centers that research will specifically consider performance variables of interest to industry would be rightly viewed as a degradation of the university research atmosphere. The suspicion that researchers whose work proved complementary to industrial research and development might expect superior salary and promotion opportunities could jeopardize even less objectionable efforts to increase the output of such work.

Finally, communication between firms and university researchers is an important criterion whose satisfaction cannot be taken for granted. If a firm contracts with academic researchers for a particular investigation, communication of the results to users within the firm is easily arranged. But when firms become interested in a broader range of research activities being conducted on a continuous basis at MRCs, users will not necessarily receive relevant information.

A Survey of American Microelectronics Research Centers

There is thus no guarantee that universities will satisfy vertical criteria in research on microelectronics. To accomplish that will require considerable effort, patience, and creativity in institutional design. Because administrators and faculty at research centers cannot be expected to specify the conduct of experimental research in contracts, an atmosphere of open communication and trust is necessary if firms are to support research programs on a continuing basis. Responsiveness to industrial needs can be achieved, not through administrators' promises or policies, but rather by providing firms with opportunities to stimulate genuine interest among academicians in specific problems in microelectronics technology. Universities with nationally recognized research programs and a history of extensive industrial contacts are in the strongest position to develop such creative interactions, because personal relationships have already been established and researchers know each other's capabilities and needs. This, in fact, is the typical background to cooperative industry support of MRCs.

The Microelectronics Center of North Carolina (MCNC), by contrast, has been established with major funding from the state (starting in 1981 with a two-year grant of $24.4 million) and a minimum of corporate support. Although founded amid expectations that it would attract microelectronics firms to North Carolina, MCNC's first priority is the improvement of microelectronics-related programs at the institutions it serves: the University of North Carolina at Chapel Hill, the University of North Carolina at Charlotte, North Carolina State University at Raleigh, North Carolina A&T University at Greensboro, Duke University, and the Research Triangle Institute. MCNC promotes these programs by supplementing the funds available for faculty salaries and graduate student fellowships, by developing curricula, and, most dramatically, by sponsoring a centralized research laboratory with microelectronics processing capabilities comparable to those in industrial facilities. The priority given to industrial-style laboratory facilities enables MCNC to exploit industrial contacts that will occur as university programs become more dynamic.[51] At other MRCs corporate participants have contributed to the establishment of comparable facilities, so MCNC may indeed attract the kind of industrial involvement present at other universities.

Table 2 summarizes the different forms that corporate involvement has taken in MRCs around the United States.[52] At Berkeley and MIT, efforts to improve microelectronics facilities have been accompanied by a minimum of internal change: research policy remains the province of departments and interdepartmental working groups. Berkeley, which has begun to receive substantial assistance from the state for improving its research equipment, receives feedback from the private sector through its long-established Industrial Liaison Program, which involves some seventy firms. Academicians and corporate researchers alike rotate between university and private laboratories and provide each other assistance in experimental research.[53] MIT, with its ambitious plan for improving physical facilities, has established a Microsystems Advisory Board through which corporate donors can channel advice.[54] Like Berkeley, it maintains a corporate liaison program involving staff from numerous firms. Stanford's new Center for Integrated Systems can employ up to 30 faculty members and 250 doctoral candidates, all of whom will retain affiliations with regular university departments. Corporate sponsors serve on an advisory committee, and corporate staff are given office space at CIS, conduct seminars, and have the opportunity to work on research projects. One of the codirectors has remarked that CIS presents an unusual opportunity for graduate students to come into contact with representatives of the industrial world and the current frontiers of industrial technology.[55]

In defining objectives for their microelectronics research activities, Stanford, MCNC, and to a lesser extent MIT have emphasized the acquisition of integrated circuit processing capabilities comparable to those used in industry. In fact, when faculty expertise in developing the software and instrumentation needed for computer-controlled processing is present, such university production lines may be close to state-of-the-art—in contrast to the more accessible research equipment permitting extensive "hands-on" experience that is characteristic in other universities (and still a feature of Berkeley's new laboratory facilities).[56] This illustrates how a greater concern with vertical aspects

Table 2. U.S. Microelectronics Research Centers Featuring Ties to Firms

---

Major Attempts to Achieve New Recognition in Field

- North Carolina (6 research institutions)

  Microelectronics Center of North Carolina established by state in 1981 to strengthen research capabilities of five universities. Includes IC manufacturing facility. To date, chiefly a state effort; General Electric has been influential behind the scenes.

- Arizona State University

  Center for Engineering Excellence established ca. 1981 to greatly expand scope and scale of engineering programs. Major support from state and industry.

Preexisting Leaders in Field

- Berkeley

  MRC established 1960. Industrial Liaison Program. Research equipment to be upgraded through major new state funding; industrial sponsors to share research costs. No new organizational arrangements.

- Stanford

  Center for Integrated Systems established 1981 with strong industry support. Industrial Affiliates Program. Ultramodern IC manufacturing facility under construction.

- MIT

  Microsystems Industrial Group established ca. 1981 to create ultramodern IC manufacturing facility, with funds coming from government research overhead and industry. Microsystems Advisory Council to advise on programs. Numerous microelectronics-related activities; no designated center to centralize research policies.

Table 2, continued

Emphasis on Creating Interdisciplinary Centers

- Cornell

  National Research and Resource Facility for Submicron Structures established under NSF auspices in 1977. Emphasis on developing exceptional facilities for experimentation to be used by outside researchers. Industrial affiliates program.

- Rensselaer

  Center for Integrated Electronics established ca. 1981 with academic focus. Costs covered by industry.

- University of Minnesota

  Microelectronics and Information Science Center established ca. 1980. Local firms provide major funding and are involved in MEIS governance. Firms may contribute adjunct professors.

Specialized Focus

- California Institute of Technology

  Silicon Structures Project established ca. 1978 to develop integrated approach to CAD. Firms contribute $100,000 and send researcher for twelve-month sabbatical. Prominent and controversial approach to CAD based on work of Mean and Conway.

---

of microelectronics technology can alter the research atmosphere at universities.

The new processing centers at Berkeley, MIT, Stanford, and MCNC are university facilities. Although this is convenient for academic researchers and tends to preserve an atmosphere of academic autonomy, as well as contributing to the prestige of the institutions owning the facilities, substantial costs are incurred through operations and obsolescence. Contracting with commercial facilities to have experimental devices fabricated would be a less costly alternative. A different cost-saving option, adopted by the University of Minnesota's Microelectronics and Information Sciences Center (MEIS), allows university researchers access to the processing facilities of nearby corporate contributors, including Control Data, Honeywell, Sperry, and 3M.

Of the MRCs described here, MEIS probably exhibits the greatest degree of corporate participation in administrative direction. Sponsoring firms send representatives to its board of directors, and company researchers can serve as adjunct professors.[57] This is certainly consistent with Control Data's public pronouncements concerning the need for small firms (relative to IBM and the largest Japanese electronics firms) to invest jointly in experimental research rather than bearing the costs of duplicative programs.[58]

Rensselaer Polytechnic Institute and Cornell, like the University of Minnesota, enjoyed preexisting strengths in specialized fields but lacked broad-based, multidisciplinary programs for responding to industrial research needs. Renssalaer has now raised funds from industry, and Cornell from the National Science Foundation, to establish well-equipped MRCs. Corporate involvement in policy-making at these centers is less evident than at MEIS.

The Silicon Structures Project at the California Institute of Technology offers a strikingly different example of corporate involvement in academic research. Each corporate sponsor contributes $100,000 annually to the Computer Science Department and sends a researcher to spend a year in residence at the university. Faculty and students develop research projects, to which the visiting researchers attach themselves. The Silicon Structures Project is focused on a specific approach to CAD; there are no plans for developing a broad-based approach to problems in microelectronics generally.[59]

Despite these differences in the organization of MRCs, two consistent patterns are evident. First, support is multilateral, rather than the bilateral arrangements common in the bioengineering industry, where individual firms have contracted with universities to fund research centers. This multilateral involvement of firms complements the horizontal nature of the academic "idea stream": firms can exchange information concerning their approaches to innovations in input, rather than merely sponsoring research designed to support internal strategies for technological development. Second, MRCs frequently provide opportunities for corporate and academic researchers to work together. This would appear to be an efficient means of satisfying certain vertical criteria. Arguably, these contacts at the working level make academic researchers more aware of particular variables that are critical in determining technological performance. Equally important, corporate researchers gain a sustained exposure to academic research activity, which virtually assures transfer of information back to industry.

Making university research more responsive to industrial needs thus does not require that firms secure prior commitments concerning research policies or that they become involved in administrative decisions. Even at MEIS, which exhibits the greatest degree of corporate involvement, decisions concerning research projects rest with researchers, who are expected to reach a consensus among themselves.[60] This suggests that there is some cause for optimism concerning the ability of MRCs to be responsive to vertical criteria without excessive sensitivity to the desires of corporate users that would jeopardize academic freedom.

Of course, interactions at the working level do not guarantee that academic researchers will not choose projects that are, from an industrial point of view, too "academic." Providing modern equipment and creating institutions that encourage interdisciplinary collaboration may be necessary for promoting academic research that is relevant to industry, but these steps are not in themselves sufficient to insure the actual achievement. It remains to be seen whether academic researchers will find sufficient inspiration in the problems of industry to be willing to change somewhat their attitude toward microelectronics research. It is noteworthy, however, that commentary on the research frontiers of industrial technology leaves a strong impression that gaps in the industrial knowledge base can present interesting research problems to academic scientists and engineers.[61]

Notes

1. The literature on collective action identifies a clear correlation between moves toward cooperation and the perception of a threat to a group's joint interest. See Francis W. Wolek, *The Role of Consortia in the National Research and Development Effort* (Philadelphia: University City Science Center, 1977), 30–32.

2. Jay H. Harris et al., "The Government Role in VLSI," in *VLSI Electronics: Microstructure Science*, ed. Norman G. Einspruch, 2 vols. (New York: Academic Press, 1981), 1:265–99.

3. Semiconductor Industry Association (SIA), *The International Microelectronic Challenge* (Cupertino, Calif.: SIA, 1981), 27.

4. See William C. Norris, "How to Expand R&D Cooperation," *Business Week*, 11 April 1983, 21.

5. The distinction between complements to, and substitutes for, internal research and development is from David C. Mowery, "The Nature of the Firm and the Organization of Research: An Investigation of the Relationship between Contract and In-House Research" (Working paper, Division of Research, Graduate School of Business Administration, Harvard University, 1982).

6. Richard C. Levin, "Innovation in the Semiconductor Industry," in *Govern-*

ment and Technical Change: A Cross-Industry Analysis, ed. R. R. Nelson (New York: Pergamon Press, 1982), 50.

7. Carver Mead and Lynn Conway, *Introduction to VLSI Systems* (Reading, Mass.: Addison Wesley, 1980), 38.

8. Arthur L. Robinson, "New Ways to Make Microcircuits Smaller," *Science*, 30 May 1980, 1019.

9. R. W. Keyes, quoted in Robinson, "New Ways to Make Microcircuits Smaller."

10. Several leading firms are involved in improvements in the process of producing crystalline silicon. See U.S. Department of Commerce, Industry and Trade Administration, Office of Producer Goods, *A Report on the U.S. Semiconductor Industry* (Washington, D.C., 1979), 19. New integrated circuit fabrication equipment from outside suppliers is often an outcome of research and development conducted by microelectronics producers. See Eric von Hippel, "The Dominant Role of the User in Semiconductor and Electronic Subassembly Process Innovation," *IEEE Transactions on Engineering Management* EM–24, no. 2 (May 1977): 60–71.

11. See Nico Hazewindus, *The Microelectronics Capability of the United States* (New York: New York University Center for Science and Technology Policy, 1981), 145; and Robinson, "New Ways to Make Microcircuits Smaller," 1022.

12. See Robinson, "New Ways to Make Microcircuits Smaller." William I. Strauss also presents this view in *Status '81: A Report on the Integrated Circuit Industry* (Scottsdale, Ariz.: Integrated Circuit Engineering Corporation), 85: "In the past the U.S. has been the primary source of IC technology, and today's products manufactured in Europe and Japan still use U.S. basic processes such as silicon-gate NMOS. There is considerable concern about tomorrow, as the U.S. merchant suppliers are contributing very little to basic research. . . . In spite of the spectacular progress in integrated circuit technology in the last decade, there have been comparatively few unique process inventions during the period. Almost all of the successful structures were at least envisioned in the late sixties."

13. Almarin Phillips, "Organizational Factors in Research and Development and Technological Change: Market Failure Considerations," in *Research, Development and Technological Innovation*, ed. Devendra Sahal (Lexington, Mass.: Lexington Books, 1980), 114–15.

14. The concept of bottlenecks in a technological system is discussed in general terms in Devendra Sahal, "Alternative Conceptions of Technology," *Research Policy* 10 (1981): 12–17, and Karol I. Pelc, "Remarks on the Formulation of Technology Strategy," in *Research, Development and Technological Innovation*, ed. Devendra Sahal (Lexington, Mass.: Lexington Books, 1980), 232–33.

15. R. W. Keyes, "Limitations of Small Devices and Large Systems," in *VLSI Electronics: Microstructure Science*, ed. Norman G. Einspruch, 2 vols. (New York: Academic Press, 1981), 1:188.

16. National Research Council, Panel on Thin-Film Microstructure Science

and Technology, Solid State Sciences Committee, Assembly of Mathematical and Physical Sciences, *Microstructure Science, Engineering and Technology* (Washington, D.C.: National Academy of Sciences, 1979), 7.

17. Strauss, *Status '81*, 85.

18. Mel H. Eklund and William I. Strauss, ed., *Status '82: A Report on the Integrated Circuit Industry* (Scottsdale, Ariz.: Integrated Circuit Engineering Corporation, 1982), 65–66.

19. Hazewindus, *Microelectronics Capability*, 174–87.

20. A. Phillips and B. Katz, "Government, Technological Opportunities, and the Structuring of the Computer Industry," in *Government and Technical Change*, ed. R. R. Nelson (New York: Pergamon Press, 1982), 40.

21. Strauss, *Status '81*, 43; Arthur L. Robinson, "Are VLSI Microcircuits Too Hard to Design?" *Science*, 11 July 1980, 262.

22. "Hewlett-Packard's New Desktop Model of Minicomputer Has Mainframe Power," *Wall Street Journal*, 17 November 1982, 8.

23. "Intel's Biggest Shrinking Job Yet," *Fortune*, 3 May 1982, 250–56. The original design envisioned 230,000 transistors on the three chips.

24. Hazewindus, *Microelectronics Capability*, 146–47.

25. Hazewindus, *Microelectronics Capability*, 151.

26. Hazewindus, *Microelectronics Capability*, 152–53; Robinson, "Are VLSI Microcircuits Too Hard to Design?"

27. National Research Council, *Microstructure Science*, 7.

28. National Research Council, *Microstructure Science*, 7.

29. Robinson, "New Ways to Make Microcircuits Smaller," 1022; Hazewindus, *Microelectronics Capability*, 145.

30. William I. Strauss, ed., *Status '83: A Report on the Integrated Circuit Industry* (Scottsdale, Ariz.: Integrated Circuit Engineering Corporation, 1983), 29.

31. Keyes, "Limitations of Small Devices and Large Systems," 187–88.

32. National Research Council, *Microstructure Science*.

33. M. P. Lepselter and W. T. Lynch, "Resolution Limitations for Submicron Lithography," in *VLSI Electronics: Microstructure Science*, ed. Norman G. Einspruch, 2 vols. (New York: Academic Press, 1981), 1:84.

34. "Semiconductor Makers Facing Costly Choice on Equipment," *Wall Street Journal*, 15 October 1982, 37.

35. W. F. Brinkman, "Microfabrication and Basic Research," in *VLSI Electronics: Microstructure Science*, ed. Norman G. Einspruch, 2 vols. (New York: Academic Press, 1981), 2:160.

36. National Research Council, *Microelectronics Science*, 19–20.

37. Fred W. Voltmer, "Manufacturing Process Technology in MOS VLSI," in *VLSI Electronics: Microstructure Science*, ed. Norman G. Einspruch, 2 vols. (New York: Academic Press, 1981), 1:13–14. Among the techniques under development are plasma, reactive ion, sputter, and ion milling approaches to etching.

38. Voltmer, "Manufacturing Process Technology," 14–22. Techniques include atmospheric-pressure chemical vapor deposition (CPD), low-pressure CPD, and plasma-assisted CPD. Possible materials include silicides, silicide/polysilicon composites, and composites of different metals with silicides or polysilicon.

39. Strauss, *Status '83*, 3. Projections for 1981–85 anticipated demand to nearly double in dollar terms.

40. Eklund and Strauss, *Status '82*, 65.

41. Voltmer, "Manufacturing Process Technology," 32–34.

42. Hazewindus, *Microelectronics Capability*, 150.

43. See Phillips, "Organizational Factors," 110–20.

44. For a similar comparison of institutional efficiency in a different technological context, see Oliver E. Williamson, "The Organization of Work: A Comparative Institutional Assessment," *Journal of Economic Behavior and Organization* 1 (1980): 5–37.

45. See "Atom by Atom, Physicists Create Matter that Nature Has Never Known Before," *New York Times*, 1 June 1982, C1.

46. D. C. Gazis, "Influence of Technology on Science: A Comment on Some Experiences at IBM Research," *Research Policy* 8 (1979): 244–59.

47. Eklund and Strauss, *Status '82*, 52.

48. Ernest Braun and Stuart MacDonald, *Revolution in Miniature: The History and Impact of Semiconductor Electronics* (Cambridge: Cambridge University Press, 1978), 127.

49. Hazewindus, *Microelectronics Capability*, 197.

50. Hazewindus, *Microelectronics Capability*, xi.

51. William Cromie, "University–Industry–State Government Consortia in Microelectronics Research," in National Research Council, *Microstructure Science*, 248–49.

52. Material reported in this section was current as of 1982.

53. Dave Hodges (University of California, Berkeley), remarks at "Cooperation and Sharing among U.S. Microelectronics Centers," a workshop sponsored by the National Science Foundation, Washington, D.C., 6 April 1982.

54. Hazewindus, *Microelectronics Capability*, 181.

55. John Linvill, quoted in "Looking Ahead to the Silicon Future," *New York Times*, 1 December 1981.

56. Hazewindus, *Microelectronics Capability*, 185–86.

57. "Joining Hands against Japan," *Business Week*, 10 October 1980, 108–13; Robert Hexter (University of Minnesota), remarks at "The U.S. Microelectronics Industry and Public Policy," a seminar at New York University for Science and Technology Policy, 2 April 1982.

58. "U.S. Electronics Firms Consider Joining in Research Venture to Counter Japanese," *Wall Street Journal*, 1 March 1982, 6; "Control Data Seeks to Form VLSI R&D, Production Cooperative," *Electronic News*, 13 April 1981, 1.

59. George Lewicki (California Institute of Technology), remarks at "Coop-

eration and Sharing among U.S. Microelectronics Centers," workshop sponsored by the National Science Foundation, Washington, D.C., 6 April 1982.

60. Hexter, remarks at "U.S. Microelectronics" seminar.

61. See Robinson, "New Ways to Make Microcircuits Smaller," and idem, "Are VLSI Microcircuits Too Hard to Design?"

Part Two

Location of High-Technology Industry: The Regional Perspective

Factors Affecting Manufacturing Location in North Carolina and the South Atlantic

John S. Hekman and
Rosalind Greenstein

For decades, manufacturing employment in the United States has grown faster outside than inside the traditional manufacturing belt;[1] this trend has accelerated in recent years. Causes of this redistribution may include surplus farm labor in the South,[2] an attractive climate in the West, and unionization and high taxes in the North. Changes in product and process technology have also had dramatic effects on the location of some industries.

Manufacturing has employed a steadily declining share of the American labor force in the twentieth century. As the wages of production workers have risen about 2 percent per year after inflation for most of the century, industry has adopted mass production technologies that replace the high-wage skilled workers of the past with lower-skilled workers and automated machinery. As a result the number of workers in production jobs in the United States has only increased from 12 million to 14 million since 1950, while the total labor force has grown from 60 million to 100 million. Today manufacturing is a relatively stagnant sector of the labor force in comparison to the growth in services, finance, insurance, real estate, and other areas.

But what is true for the nation as a whole is not true for individual regions. The changing nature of manufacturing technology and also the changing distribution of population and markets have resulted in a dramatic shift in the location of manufacturing employment. During the 1970s, the Manufacturing Belt lost about 1 million manufacturing jobs, while the rest of the country gained almost 2 million.[3] North Carolina and the rest of the Southeast have shared in this growth. Until the 1960s the growth of industry in the Southeast was predominantly in textiles, apparel, furniture, and forest products. As a rule this growth was mainly in nondurables and "light" as opposed to "heavy" industry. The region had few industrial complexes, where related industries fed off each other's growth (such as steel, metalworking, machine tools, and durable goods producers), like those in the Northeast.

In the 1970s, however, a wide range of industries began to locate in the Southeast. Table 1 shows the growth and decline of manufacturing in the Southeast during that decade. The figures presented demonstrate clearly that a complete turnaround occurred in the nature of the region's industrial growth. The fastest-growing industries were rubber and plastics, fabricated metals, machinery, electronics, and instruments. The traditional southern industries—textiles, apparel, wood products, furniture, and paper—all grew more slowly or declined.

North Carolina's turnaround was not so pronounced as that for the region as a whole. While textile employment declined much more in the state than in the region, the performance of other industries was more balanced between the old and the new. Apparel, wood products, furniture, chemicals, and paper grew more rapidly in North Carolina than in the region; machinery, electronics, and instruments grew more slowly. Electronics and instruments are often included in the designation "high technology." Florida recorded the largest absolute gains in these two industries. North Carolina was second in the number of jobs gained in these fields but ranked quite low in percentage terms.

Most studies of industry location concentrate on regional differences in production costs and on the location of markets for inputs and final products. This kind of analysis is less relevant today, especially for electronics and other high-technology industries. These industries typically have high value-to-weight ratios; thus transportation cost is not a significant share of production cost, and choice of site is not tied to the location of natural resources or energy supplies. These factors make the newer industries footloose, searching out low-wage labor.

The footloose nature of many industries today has resulted in an increasing dispersion in the location of manufacturing facilities, while corporate headquarters have become if anything more concentrated in major urban centers. The pattern of dispersion of plants over time from their original region of the country to new locations is most often analyzed by using life-cycle models of products and production processes. The product life-cycle model of industry growth and location relies on the distinction between *specialized* and *ubiquitous* inputs.[4] Some early part of the design or manufacturing of a product requires highly *specialized* resources, because of rapid change in technology. In this stage, firms are most productive in large urban agglomerations where a wide range of skills and resources can be tapped.[5] Agglomeration of industry results from both interdependence among firms and reliance on a common labor pool. Higher productivity offsets the higher costs of labor, land, and other factors that result from this concentration.[6] The resources that attract an industry to an agglomeration are specialized and have generally been available only in the urban industrial centers. For a product sold in the national market, the production processes that can be performed with *ubiquitous* inputs (conventional labor and capital) will be located outside the industrial center, because such resources cost less there. Conversely, production processes that require specialized resources occur in the agglomerations where these resources are available; these processes use the minimum necessary amount of ubiquitous inputs (because their prices are higher there).

In the early stage of the product life cycle, production tends to be small-scale and needs a large amount of scientific and engineering support. Frequent changes in product design mean that production runs are short, so that mass production is inefficient. These changes also mean that the design and production facilities will both be located in the home region, because frequent, nonroutine communication between the two is essential. This orientation to the home region and specialized inputs changes over time as the production process and production design become sufficiently standardized to allow the substitution of unskilled labor and automated production techniques for low-volume, skilled assembly work. The size of the establishment increases to take advantage of mass production, and production processes spread out geographically toward the lowest-cost locations for their respective bundles of inputs. In this way, a component that needs primarily low-skilled labor can be produced overseas in a low-wage country, and subassemblies requiring extensive metalworking can be produced in a heavy-industry region, while design and development work normally remains in the home region because of its skill inputs and closer interaction with the direction of the enterprise.

The redistribution of manufacturing that has been taking place in the United States for the past several decades exhibits the basic characteristics of this product cycle model. While the control of manufacturing as measured by headquarters location and research and development activities has become more centralized,[7] production facilities have become more decentralized. During the 1970s 75 percent of new plants built by Fortune 500 firms were located outside the core manufacturing region.[8] For North Carolina, this has meant that new production facilities are dominated by branch plants of firms whose headquarters are outside the state, mostly in the Northeast and the Midwest. Because research and development are for the most part performed at or near the headquarters facilities of most firms,[9] the new industrial plants sited in North Carolina, even those in high-technology industries, are mainly used for production rather than design or central

Table 1. Employment Gains/Losses by Industry in the Southeast, 1968-1978, in thousands (with % change)

Industry	AL	FL	GA	LA
Textile mill products	-0.7 (-1.8)	2.1 (84.0)	3.1 (2.7)	2.7 (*)
Apparel	11.8 (28.6)	16.2 (91.5)	8.9 (13.1)	3.6 (47.4)
Lumber, wood products	2.0 (10.0)	6.3 (49.2)	9.0 (51.7)	-1.4 (-9.2)
Furniture, fixtures	3.0 (68.2)	2.6 (37.7)	0.7 (7.4)	-13.4 (-92.4)
Paper	2.2 (13.9)	-0.4 (-2.4)	3.8 (16.4)	8.7 (145.0)
Chemicals	0.6 (4.8)	1.1 (5.3)	3.1 (25.0)	10.0 (47.6)
Rubber, plastic products	6.7 (80.7)	7.4 (148.0)	8.1 (168.8)	-0.4 (-100.0)
Fabricated metal products	9.8 (61.3)	8.7 (45.8)	6.1 (44.5)	4.6 (47.4)
Machinery except electrical	6.1 (65.6)	10.2 (68.0)	7.1 (62.8)	5.4 (110.2)
Electronic equipment	9.8 (118.1)	24.7 (93.6)	8.5 (87.6)	7.1 (191.9)
Transportation equipment	-1.6 (-7.9)	9.3 (32.2)	-10.0 (-22.9)	8.3 (51.9)
Instruments	2.5 (625.0)	7.2 (240.0)	2.9 (145.0)	-0.5 (-100.0)
TOTAL MANUFACTURING	48.8 (16.4)	101.7 (34.5)	82.6 (19.3)	31.1 (18.3)

*Infinite percentage gain

MS	NC	SC	TN	VA	TOTAL
0.0 (0.0)	-26.5 (-9.8)	-6.3 (-4.5)	-5.1 (-15.3)	3.3 (8.1)	-27.4 (-4.2)
1.7 (4.8)	10.8 (16.1)	2.1 (4.5)	5.1 (7.7)	5.4 (17.4)	65.6 (17.2)
1.1 (5.3)	8.4 (32.2)	1.6 (12.2)	3.1 (21.7)	2.9 (15.3)	33.0 (21.1)
3.0 (25.2)	22.4 (37.1)	-0.9 (-17.6)	1.2 (5.5)	2.3 (9.3)	20.9 (31.1)
0.5 (8.1)	3.4 (21.8)	1.3 (11.8)	2.8 (20.4)	-0.2 (-1.5)	22.1 (18.2)
1.7 (34.7)	13.2 (68.0)	12.3 (56.9)	-1.2 (-2.2)	-10.8 (-24.9)	30.0 (14.3)
-2.4 (-54.5)	15.6 (177.3)	10.4 (315.2)	13.6 (136.0)	5.0 (71.4)	64.0 (123.1)
2.2 (25.9)	14.6 (101.4)	8.2 (124.2)	10.7 (42.6)	4.1 (33.3)	69.0 (55.1)
5.0 (58.8)	11.5 (49.4)	16.5 (123.1)	14.2 (79.8)	8.9 (95.7)	84.9 (75.3)
9.5 (86.4)	10.3 (30.8)	6.5 (49.2)	9.9 (32.8)	3.7 (15.0)	90.0 (56.1)
13.8 (97.9)	7.0 (14.8)	-0.8 (-23.5)	8.7 (48.3)	6.8 (22.8)	41.5 (23.0)
0.1 (4.3)	3.1 (54.3)	3.1 (54.3)	2.5 (113.6)	3.1 (140.9)	24.0 (114.8)
52.5 (31.0)	120.0 (17.9)	67.4 (21.1)	21.1 (75.8)	51.0 (14.4)	630.9 (20.1

administration. The results of the North Carolina industrial location study and the high-technology location study reported below confirm this picture of the nature of the state's newer industries.

South Atlantic Site Selection Survey

In 1981, 317 industrial facilities that had opened or been expanded since 1977 in Virginia, North Carolina, and South Carolina were surveyed to study the reasons for their choice of a site.[10] In a combination of mail and telephone interviews, 204 responses were recorded.[11] The decision-makers were asked to rate, on a scale of 1 to 5, the importance of nineteen business factors and twelve quality of life factors in their choice of a site. Because the study sought to discriminate between what they looked for and what they actually found, respondents were also asked to judge whether the attributes of the site they selected were better than, the same as, or worse than their next best alternative. Unlike other studies of this kind, this one asked firms to identify other possible sites that had been considered for the facility. Knowing the identity of these alternative sites permits a comparison of states in direct competition for industry.

The Key Factors in the Location Decision

Industrial site selection is complicated not only by the myriad economic and social factors that can vary widely within and between regions, but also by the numerous layers of management responsibility in many companies. Most facilities in this survey were owned by corporations that operated plants in more than one state. As a result they had as many as three levels of management involved in the site selection process: national headquarters, regional or divisional headquarters, and plant management.

Most surveys find that small firms, especially those with only one establishment, have a much narrower focus for their site selection than large companies with plants in several states. Most of the small firms consider sites in only one state, and a large percentage look at just one site, usually close to the owner's place of residence. Large firms, on the other hand, are becoming increasingly sophisticated at compiling data on many sites and making the selection on the basis of many characteristics. In general, the greater the distance a firm's headquarters is from a particular state, the greater the competition that state faces in attracting that industry.

Table 2. North Carolina Facilities by Type of Growth and Headquarters Location

Location of Headquarters	Type of Growth				
	Independent[a]	New Branch	Expansion	Relocation	TOTAL
In state	16	15	7	4	42
Out of state	2	53	13	6	74
TOTAL	18	68	20	10	116

[a]Single-establishment firm

An important characteristic of the facilities surveyed for this report (selected in random fashion) is that the great majority were headquartered out-of-state. In North Carolina, most of the plants were new branches rather than headquarters (59 percent); headquarters for these branches are owned out-of-state in 78 percent of the cases, and 64 percent of all plants are out-of-state owned (table 2).

The survey also asked explicitly where the site location decision had been made, at the plant level or at regional or national headquarters. Table 3 shows the number of decisions made at the different levels for the 116 North Carolina facilities that responded to this question. In some cases respondents indicated that several management levels were involved in the decision, but the basic decision was most often made by national headquarters. Our survey indicates that the decision to locate a facility in North Carolina was most often made in another state. Other studies, such as that by David Birch,[12] have also found that a large fraction of new manufacturing in the Sunbelt is composed of branch plants of northern firms.

The distribution of headquarters states for these facilities is quite revealing. As table 4 shows, 80 percent of the new branch plants in North Carolina are operated by firms from the Northeast and North Central regions. New York, Michigan, and Ohio together account for 42 percent of the branches. In some cases, especially New York and Illinois, this pattern may mean only that corporate headquarters is located in New York City or Chicago, with manufacturing capacity entirely in the Southeast. But more often the headquarters–branch

Table 3. Where the Site Decision Was Made for North Carolina Facilities

Location of Decision-Maker	Frequency	%
Plant personnel	28	24
Regional headquarters	14	12
National headquarters	72	62
Other	2	2
TOTAL	116	100

pattern in this survey represents a decision to locate new capacity in the Southeast, breaking with the old tradition of expanding within the Manufacturing Belt.

Which Industries Are Coming

Table 5 presents the industry distribution for the new and expanded facilities in the survey. Because the individual plants were selected randomly from published lists of announced openings and expansions, this distribution can be considered fairly representative of the pattern of industrial growth in these three states. New employment was created in all twenty industry groups except leather goods. Even textiles, where employment has been falling in recent years, is well represented; however, its percentage of representation in the sample of new plants is far less than its share of the region's economy, reflecting its shrinking importance. All of the traditional industries of the South Atlantic—tobacco, textiles, apparel, lumber, and furniture—have a smaller share in the sample than their share of the economy. Industries new to the region are well represented. Fabricated metals, nonelectrical machinery, electrical equipment, and transportation equipment together make up over 40 percent of the facilities studied.

Business Location Factors

Table 6 summarizes the ranking of nineteen business location factors for facilities in the study. The rankings are based on the mean of the

Table 4. Headquarters State for New Branch Plants, Plant Expansions, and Plant Relocations in North Carolina, 1977-1981[a]

Headquarters	Frequency	%
Mountain States	4	6
Arizona	1	1
Idaho	1	1
Wyoming	2	3
North Central States	29	43
Iowa	1	1
Illinois	4	6
Indiana	2	3
Kansas	2	3
Michigan	9	13
Missouri	2	3
Ohio	9	13
Northeast	25	37
New England:	8	12
Connecticut	4	6
Massachusetts	4	6
Middle Atlantic States:	17	25
New Jersey	3	4
New York	11	16
Pennsylvania	3	4
South Atlantic States	5	7
Georgia	2	3
Maryland	2	3
Virginia	1	1
South Central States	3	4
Mississippi	1	1
Tennessee	1	1
Texas	1	1
Foreign	1	1

[a]Excluding firms headquartered in North Carolina

Table 5. Distribution of New Plants Surveyed, by State and Industry

Industry	All	%	NC	%	SC	%	VA	%
Food	11	5.5	5	4.7	1	1.8	5	17.2
Tobacco	4	2.0	3	2.8	0	0.0	1	3.5
Textiles	16	8.0	10	9.4	4	7.1	2	6.9
Apparel	13	7.0	7	6.5	5	8.9	1	3.5
Lumber	6	3.0	5	4.7	1	1.8	0	0.0
Furniture	10	5.0	8	7.5	2	3.6	0	0.0
Paper	3	1.0	3	2.8	0	0.0	0	0.0
Printing	4	2.0	1	0.9	1	1.8	2	6.9
Chemicals	16	8.0	9	8.4	5	8.9	2	6.9
Petroleum	1	0.5	1	0.9	0	0.0	0	0.0
Rubber/plastic	9	4.5	3	2.8	6	10.7	0	0.0
Leather	0	0.0	0	0.0	0	0.0		
Stone, clay	4	2.0	1	0.9	1	1.8	2	6.9
Primary metals	4	2.0	0	0.0	4	7.1	0	0.0
Fabricated metals	16	8.0	9	8.4	6	10.7	1	3.5
Nonelectrical machinery	33	17.0	18	16.8	12	21.4	3	10.3
Electrical equipment	21	12.0	12	11.2	4	7.1	5	17.2
Transportation equipment	11	6.0	8	7.5	0	0.0	3	10.3
Instruments	4	2.0	1	0.9	2	3.6	1	3.5
Miscellaneous	4	2.0	2	1.9	1	1.8	1	3.5
TOTAL	194	100.0	107	100.0	56	100.0	29	100.0

rating (on a scale of 1 to 5) given to each factor by the managers involved in choosing the plant locations. The top five factors for all firms were (1) state and local industrial climate; (2) labor productivity; (3) transportation; (4) land availability and room for expansion; and (5) cost of land and construction. These results differ markedly from those of a similar location survey conducted by *Fortune* magazine in 1977.[13] Though business climate ranked only ninth in the *Fortune* study, it was a strong first in the South Atlantic. Nationally, firms said they were most concerned with transportation and proximity to customers, but in this region transportation stood third and proximity to markets only eleventh. Land availability and the cost of land and construction were likewise much more important in the South Atlantic. The best generalization of these results for the South Atlantic as compared to the whole country is that firms come to this region for its lower overall production costs—from labor productivity, land, and construction—and for its business climate. Nationally, on the other hand, most firms are more market-oriented, seeking good transportation and proximity to markets.

The similarities between the two studies are also revealing. In both surveys, state financial inducements and environmental regulations ranked relatively low. Since those are the factors most under the control of state economic development agencies, these results have led some to suggest that plants are not lured to particular states by industrial recruiters so much as they are drawn by overall economic factors such as labor rates. In the telephone interviews, fewer than a half-dozen firms said they had been influenced substantially by the states' financial programs.

The Influence of Quality of Life

In addition to business factors as influences on the choice of a site, firms were asked to rate the importance of factors representing the quality of life. Because manufacturing today is tied less to specific locations for natural resources, rail connections, or material inputs, industry is somewhat more footloose than in the past. Thus it is possible for firms to choose a location at least partly on the basis of the amenities an area can offer to employees, such as climate, low cost of living, and the quality of public education. The respondents in the survey affirmed that quality of life did play a part in many of the location decisions, although often this issue emerged when comparing different sites in the South Atlantic region rather than, for example, North Carolina versus Colorado. Overall, quality of life factors received an average score of 2.81 (on a scale of 5), compared with 3.95 for the business factors. North Carolina facilities rated quality of life slightly higher than average, at 2.94. And large plants in North Carolina (employing 250 or more workers) gave quality of life factors an average score of 3.22. So it appears that these factors are being considered, though the responses were quite subjective.

The ranking of quality of life factors by industry is given in table 7.

Table 6. Ranking of Business Location Factors by Major Industry Groups (SIC) Surveyed

Business Location	Rank (all firms)	Textiles (22)	Apparel (23)	Furniture (24)	Chemicals (28)
State/local industrial climate	1*	2*	5*	3*	2*
Labor productivity	2*	3*	1*	2*	9
Transportation	3*	9	3*	7	1*
Land availability/ room for expansion	4*	4*	8	1*	6
Cost of land and construction	5*	1*	7	4*	3*
Wage rate	6	6	3*	6	12
Business taxation	7	13	6	7	6
Electricity availability/cost	8	12	9	12	10
Skilled labor supply	9	7	2*	4*	15
Proximity to suppliers/services	10	14	10	16	13
Proximity to markets	11	18	13	10	5*
Unskilled labor supply	12	8	15	11	14
State/local environmental regulations and permit processing	13	9	12	13	4*
Water supply	14	11	14	17	6
Availability of technical training programs	15	17	19	18	18
Fuel availability/cost	16	16	11	9	11
State financial incentives	17	15	17	14	14
Public wastewater treatment capacity	18	5*	15	14	19
Solid/hazardous waste disposal facilities	19	19	17	19	16
No. of facilities	194	14	13	10	15

*Top five factors

Rubber & Plastic (30)	Fabricated Metals (34)	Nonelectrical Machinery (35)	Electrical Equipment (36)	Transportation Equipment (37)
4*	1*	1*	2*	1*
2*	7	2*	1*	2*
3*	9	4*	8	3*
4*	4*	3*	4*	5*
11	2*	5*	10	10
4*	9	7	3*	9
9	2*	5*	11	5*
4*	6	8	9	8
9	11	11	6	4
13	11	10	13	10
1*	5*	12	15	10
8	15	16	7	13
17	17	12	12	16
14	16	15	15	16
10	8	8	5*	7
14	11	17	14	13
16	14	14	17	13
18	18	19	18	16
19	19	18	19	19
9	17	34	22	11

Table 7. Ranking of Quality of Life Factors by Major Industry Groups (SIC) Surveyed

Community Quality of Life Factors	Mean Score (all firms)	Rank (all firms)	Textiles (22)	Apparel (23)	Furniture (24)
Educational system	3.2	1*	1*	1*	2*
Cost of living	3.0	2*	3*	2*	2*
Housing	3.0	3*	2*	4*	0
Physical quality of air and water resources	2.8	4*	6	4*	4*
Personal taxes	2.8	5*	11	6	4*
Recreational opportunities	2.8	6	6	8	8
Climate	2.7	7	4*	2*	9
Transportation (local traffic conditions/public transportation)	2.6	8	5*	7	4*
Open space	2.5	9	8	11	4*
Cultural resources	2.5	10	9	8	11
Aesthetic quality of natural landscape (scenery)	2.5	11	12	8	1*
Entertainment	2.4	12	10	11	11
No. of facilities		194	14	12	10

*Top factors

Education and cost of living scored near the top for all industries; climate and scenery were not very important. Housing and personal taxes ranked higher with the industries that are moving into the South Atlantic at present than they did with the industries already concentrated there, such as textiles and apparel.

High-Technology Industrial Location

In the midst of the rapidly changing pattern of manufacturing location today many states and urban areas have made the attraction of high-

micals (28)	Rubber & Plastic (30)	Fabricated Metals (34)	Nonelectrical Machinery (35)	Electrical Equipment (36)	Transportation Equipment (37)
1*	2*	3*	1*	1*	1*
3*	1*	1*	4*	4*	4*
2*	8	4*	2*	2*	2*
3*	2*	6	8	6	8
6	5*	2*	5*	4*	9
7	11	5*	7	2*	3*
7	7	8	6	6	6
3*	10	12	3*	10	9
9	5*	9	9	10	12
11	9	6	10	9	4*
12	4*	9	11	8	11
10	12	11	12	12	6
14	9	17	34	22	10

technology industry a central concern. Areas that have declining industrial bases see high technology as a desirable way to replace lost jobs, and growing regions like the South Atlantic and the Southwest want high technology because it is perceived as a source of future growth, as environmentally "clean," and as a way to develop a more skilled labor force. In this section a case study of the computer industry is used to illustrate some of the location factors that are important for high technology, especially as they relate to states, such as North Carolina, that are outside the main high-technology industry centers.[14] Computer manufacturing is an ideal subject for high-technology location analysis because it consists of both high-technology and low-technology prod-

Table 8. Location of Establishments in the Computer Industry (SIC 3573), February 1979

Location	Single Location		Headquarters		Branch Plants		Total Plants		Total Employment as % of U.S. Total for Industry
	No.	Employees	No.	Employees	No.	Employees	No.	Employees	
States									
California	72	4,100	67	27,443	77	38,438	216	69,981	28.5
Massachusetts	31	2,585	23	11,666	26	17,226	80	31,477	12.8
New York	24	1,757	13	3,276	9	21,357	46	26,390	10.7
Minnesota	5	203	9	6,160	16	13,787	30	20,150	8.2
New Jersey	14	710	8	853	10	14,325	32	15,888	6.5
Regions									
Northeast:	85	5,904	54	18,761	66	63,256	205	87,921	35.8
New England	42	3,279	26	13,826	37	22,744	105	39,849	16.2
Middle Atlantic	43	2,625	29	4,935	29	40,512	100	48,072	19.6
North Central	39	1,901	23	12,566	45	17,233	107	31,701	12.9
South Atlantic	16	488	8	2,912	31	10,782	55	14,560	5.9
South Central	18	1,316	8	5,980	23	10,392	49	17,688	7.2
Mountain	15	1,289	4	4,706	16	15,701	35	21,969	8.8
Pacific	77	5,240	70	28,393	79	38,477	226	72,110	29.4
U.S. TOTAL	250	16,139	167	73,318	260	155,841	677	245,676	100.0

Source: Dun and Bradstreet, Dun's Market Identifiers (1979).

ucts, using both mass production and small, special-order processes. Product lines change very rapidly, and large amounts of scientific and engineering research are needed for firms to remain competitive. Because of these characteristics, production has become somewhat specialized by region, and changes in the industry's products over time have influenced where most of its employment is located.

Table 8 presents the geographic distribution of employment and number of establishments with 15 or more workers for computer manufacturing as of February 1979. Employment is quite concentrated, with the top five states accounting for about 67 percent of the total 545,676 employed in that category. About one-third of employment is in the Northeast, one-third is on the West Coast, and one-third is distributed throughout the remainder of the country.

Table 8 gives a breakdown of establishments into single-location firms, headquarters of multi-establishment firms, and branch establishments. Almost two-thirds of all employment is concentrated in branches, and when the headquarters plants are added, the proportion rises to 93.3 percent. The single-establishment firms comprise only 6.7 percent of employment and are for the most part small facilities—65 workers per plant, compared with 439 for headquarters and 599 for branches.

The predominance of multi-establishment firms in computer manufacturing indicates that it is a relatively mature industry, as opposed to an infant industry dominated by small research firms, as with genetic engineering. Computer companies have rationalized their production processes and taken advantage of mass production economies wherever possible. Branch plants most often contain specialized production processes rather than producing complete products. They show some tendency to cluster near their headquarters, but a good number of them are widely scattered across the country, reflecting their firms' search for advantages in production cost, labor supply, and, as discussed above, regional amenities. Branches are the mobile part of the industry, because the branch location decision is made deliberately, in most cases after the consideration of many possible locations. As the computer industry continues to mature and more product lines become mass produced items, the industry will tend to become more dispersed geographically.

Table 9 presents the regional distribution of branch plants and their headquarters for the 260 branches reported in table 8. The column totals in table 9 are the number of branches located in each region; the row totals are the number of branches nationwide that have their headquarters in each region. For example, the South Atlantic has 31

Table 9. Branch Establishments by Region and Headquarters Location for the Computer Industry, February 1979

Headquarters Location	New England	Middle Atlantic	East North Central	West North Central
New England	23	7	4	1
Middle Atlantic	5	13	3	4
East North Central	1	2	6	1
West North Central	6	5	3	14
South Atlantic	0	1	0	1
East South Central	0	0	0	0
West South Central	0	0	1	1
Mountain	0	0	1	0
Pacific	2	1	4	0
TOTAL	37	29	23	22

Source: Dun and Bradstreet, Dun's Market Identifiers (1979).

branches (column total), while 6 branches nationwide (row total) report having headquarters in the South Atlantic. The Northeast and North Central areas are exporters of branches; there are more branches nationwide whose headquarters are in these regions than there are branches located there. The rest of the nation (including California) is a net importer of branches; that is, there are more branches sited there than there are branches reporting their headquarters there.

The data include 13 computer manufacturing establishments from North Carolina (12 branch plants and 1 single-location firm). Total reported employment is 5,520. Of this, 4,000 are accounted for by one facility, the IBM installation in the Research Triangle Park. The location of the branches is dispersed throughout the state: 2 in Charlotte, 4 in Greensboro, 3 in Raleigh–Durham (including the Research Triangle), and 3 in nonmetropolitan areas. The headquarters for these branches are in Massachusetts (4), New York (3), Michigan (3), New Jersey, and Oklahoma.

The primary centers of the computer industry both export branches for some of their manufacturing processes and import branches because of the technological resources they offer. As an example of the

South Atlantic	East South Central	West South Central	Mountain	Pacific	Total
7	0	0	3	10	55
9	1	6	3	10	54
6	2	0	2	14	34
3	0	1	2	6	40
0	0	0	0	2	6
0	0	0	0	0	0
1	0	9	1	0	13
0	0	0	2	2	5
4	0	4	4	34	53
31	3	20	17	78	260

latter consideration, Minnesota firms have 6 branches in Massachusetts (primarily around Boston), and all of the major producers have facilities in California's Silicon Valley. The South Atlantic, on the other hand, is primarily a manufacturing (as opposed to research and development) center. It is the source of only 6 branches, but is the site of 31. The East North Central region, by far the largest center for all manufacturing, has very little attraction for the computer industry. It contains only 6 of the 34 branches that originate from its firms, and it is the location of only 23 branches. It ranks behind the South Atlantic in number of branches, branch employment, and total employment.

Because there has been considerable interest recently in the rapid industrial growth of the South and West relative to the North and East, the branch location pattern is shown in a "Sunbelt–Frostbelt" comparison in table 10. The Frostbelt is the source of 183 branches (70 percent), but is the home for only 111 branches (43 percent). Only 13 branches of Sunbelt firms have been located in the Frostbelt (almost all of them are clustered around the industry centers in Boston and Minneapolis). To the extent that branches represent the mobile part of the industry, the Sunbelt has received 57 percent of the branch "votes."

Table 10. Frostbelt-Sunbelt Comparison of Branch Locations in the Computer Industry, February 1979

Headquarters Location	Branch Location		Total
	Frostbelt	Sunbelt	
Frostbelt	98	85	183
Sunbelt	13	65	77
TOTAL	111	149	260

Source: Dun and Bradstreet, Dun's Market Identifiers (1979).

Another indication of how the location pattern of the industry is changing can be found by looking at where new firms are being established. Table 11 shows the age distribution of firms for the top five states by employment. Over 80 percent of firms were founded since 1960 in all five states; for California over 95 percent. From 1970 to 1979, the percentage of new firms is not too different among the top three states; California has 57 percent, New York 46 percent, and Massachusetts 43 percent. Minnesota and New Jersey are far below this. And from 1975 to 1979 the pattern looks very different; only Massachusetts is making a race of it with California.

It seems likely that the headquarters–branch pattern of computer manufacturing will not change dramatically in the near future. Currently, the largest firms are headquartered in Boston, New York, Philadelphia, Minneapolis, and California. Research and design is closely tied to the head office for all but a few of these. Four of the top five mainframe producers also have manufacturing operations near headquarters for final assembly and testing. Branch plants that have been located outside the home state or region are most often used for producing peripherals, such as terminals or printers, subassemblies, and components. Many of these branches are located in the South Atlantic and the Plains states, mostly in small to medium-sized cities. Components, such as circuit boards, that require mainly low-skilled labor are most often produced overseas in low-wage countries.

Employment should grow faster outside the industry centers than at headquarters, if only because the industry is growing and the recent trend has been to locate branches far from headquarters. How-

Table 11. Distribution of Computer Firms by Year of Founding, SIC 3573, February 1979

Industry	CA		MA		NY		MN		NJ	
	No.	%	No.	%	No.	%	No.	%	No.	%
To 1960	6	4.3	6	11.1	7	18.9	2	14.3	2	9.1
1961-65	12	8.7	3	5.5	0	0.0	1	7.1	4	18.2
1966-70	41	29.7	22	40.7	13	35.1	7	50.0	13	59.1
1971-75	44	31.9	14	25.9	14	37.8	4	28.6	1	4.5
1976-79	35	25.4	9	16.7	3	8.1	0	0.0	2	9.1
TOTAL	138	100.0	54	100.0	37	100.0	14	100.0	22	100.0

Source: Dun and Bradstreet, Dun's Market Identifiers (1979).

ever, this is not a simple production-worker–officer-worker dichotomy; computer manufacturing has a very high proportion of nonproduction workers. Firms are concerned with skilled labor supply in all of their possible sites, and even research and development activities are becoming somewhat more decentralized. The ideal location indicated by computer firms' revealed preference is the medium-sized city that combines adequate labor supply with a lower cost of living and cost of production compared to the larger headquarters city.

A Profile of North Carolina's High-Technology Sector

The existing high-technology sector in North Carolina is described below in terms of location of headquarters, size of establishment, and geographic concentration across the state. Using the Standard Industrial Classification System (SIC), high-technology industries are here considered to be electronic computing equipment (SIC 3573), electrical distributing equipment (SIC 361), electrical industrial apparatus (SIC 362), radio and television receiving equipment (SIC 365), and electronic components and accessories, including electronic tubes, capacitors, connectors, and semiconductors (SIC 367).

Data for these analyses are taken from the *Directory of North Carolina Manufacturing Firms, 1982–1983,*[15] and include 122 manufacturing establishments in nine four-digit SIC categories. Addresses of es-

Figure 1. Headquarters location of North Carolina high technology establishments, in-state locations

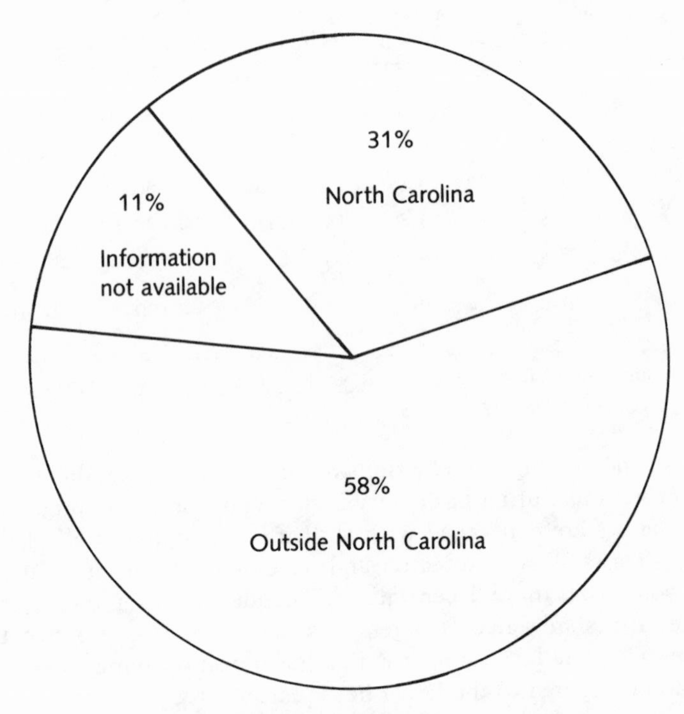

Source: *Directory of North Carolina Manufacturing Firms, 1982–1983* (Raleigh: North Carolina Department of Commerce, 1982).

tablishments and of parent firms were used to determine location; the number of employees and start-up dates were also compiled from the directory.

As shown in figure 1, only one-third of the high-technology establishments in the data base have headquarters in-state, and only one of these is a multiplant operation. Of the two-thirds of the establishments with headquarters outside North Carolina, more than 85 percent have headquarters in the Northeast or Midwest. Only eight states account for nearly all locations: New York, New Jersey, Pennsylvania, Ohio, Massachusetts, Connecticut, Indiana, and Illinois.

Like the computer industry, North Carolina's high-technology manufacturing industry as a whole has largely been created over the last two decades. Table 12 illustrates that expansion has been greater for firms with headquarters outside North Carolina. For each of the

Table 12. Start-up Date by Location of Headquarters, for Current Establishments in North Carolina's High-technology Industry

Start-up Date	Headquarters in North Carolina	Headquarters Out-of-State
1931-1940	2.0	0.0
1941-1950	1.0	2.0
1951-1960	4.0	8.0
1961-1970	11.0	19.0
1971-1981	11.0	24.0

Note: Percentages are based on 100 establishments with reported data, of 122 total surveyed.

Source: N.C. Department of Commerce, Directory of North Carolina Manufacturing Firms, 1982-1983 (Raleigh, 1982).

three decades after World War II, for all currently operating establishments, start-ups by establishments with headquarters out-of-state have accounted for nearly two-thirds of the total industry start-ups. During the 1950s, 5 of 15 high-technology establishments were classified as either single-plant operations or founded by a North Carolina firm; for the 1960s, 13 of 36 establishments were so classified, and during the 1970s, 14 of 43.

In North Carolina's high-technology industry the multilocation firm with headquarters outside the state is the predominant form of ownership. The exception is found in the computer manufacturing industry (SIC 3573), where 9 of 17 firms have headquarters inside the state. The case for the semiconductor industry is ambiguous because headquarters location was unreported for most firms. Table 13 presents these data for each industry in the study. Each firm was classified as either small (fewer than 100 employees), medium (100–249), or large (250 or more). A chi-square test revealed no significant correlation between size of establishment and location of headquarters.

Figure 2 illustrates the geographical distribution of North Carolina's high-technology employment. The greatest concentration is in the Piedmont. Within the Piedmont, Wake and Durham counties are the leaders; this reflects the importance of the roughly 20,000 high-

Table 13. Headquarters Location by SIC

SIC	Industry	NC	Out-of-State	n/a
3573	Electronic computing equipment	9	5	3
3612	Power distribution and specialty transformers	6	7	1
3613	Switchgear and switchboard apparatus	5	7	1
3621	Motors and generators	5	10	0
3622	Industrial controls	5	12	2
3661	Telephone and telegraph	1	7	0
3662	Radio and TV transmitting and detection equipment	6	8	4
3674	Semiconductors	1	1	4
3675	Electronic capacitors	0	5	0
3673	Resistors	1	5	1
	TOTAL	39	67	16

Source: N.C. Department of Commerce, Directory of North Carolina Manufacturing Firms, 1982–1983 (Raleigh, 1982).

technology manufacturing jobs in and around the Research Triangle Park. The next highest concentration of employment is in Buncombe, Catawba, Forsyth, Alamance, and Johnston counties. All of these are urbanized counties and part of a Standard Metropolitan Statistical Area, except Johnston, which is adjacent to Wake County. All counties but Wake have fewer than seven high-technology firms. Most of these counties have 500 or more employees in high-technology production firms. In addition, there are nearly 3,000 jobs near Asheville and another concentration in the three mountain counties near the Tennessee–Georgia–North Carolina border. About 1,000 jobs are located around the coastal city of Wilmington; most of these jobs are in electrical components (SIC 367). The coastal concentration of employment is in three plants, the most recent opened in 1977; the mountain region received only two plants from 1977 to 1982.

High-technology manufacturing in North Carolina thus tends to be a multibranch plant of a corporation that has headquarters outside North Carolina and outside the South Atlantic region. Size of the in-state

Figure 2. High-technology employees in North Carolina, county estimates, 1979

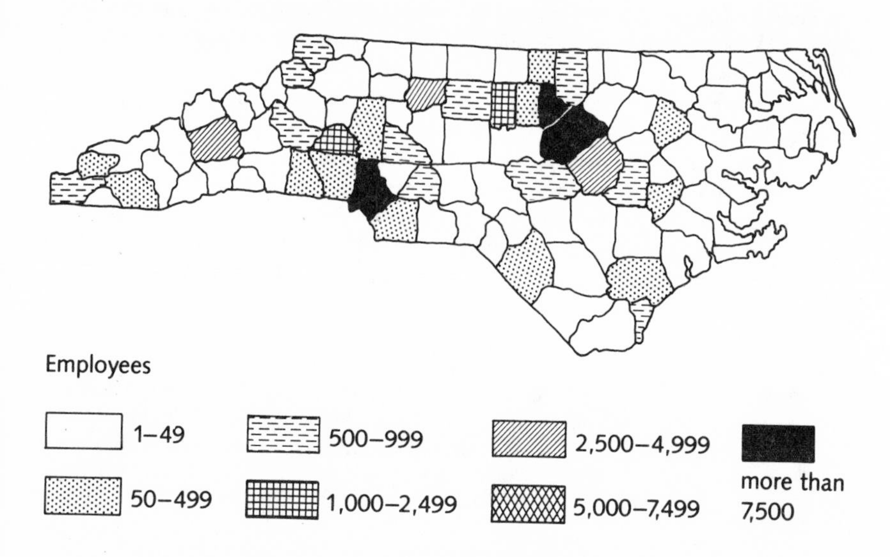

Source: U.S. Department of Commerce, Bureau of the Census, *County Business Patterns, 1979: North Carolina* (Washington, D.C., 1980).

establishment is unrelated to location of headquarters. When nationwide data are considered for the computer industry or for the 410 largest corporations, establishments of multiplant facilities with headquarters out-of-state tend to be smaller than branch plants of the same corporation located near the headquarters.[16] These results should not be confused or assumed to be contradictory; the North Carolina comparison is across corporations; the national comparison, within. Though the rate of growth for establishments headquartered out-of-state has increased in the 1980s over the 1970s, the North Carolina segment of the industry has remained stable. If this pattern continues, there will be increasingly fewer locally owned establishments. Despite the location of some installations in the border counties of the mountains and in the coastal plain, North Carolina's Piedmont is certainly the locus of the state's high-technology manufacturing. The proximity to universities and a highly skilled labor pool in the Piedmont have made it the focus of policy-makers' recruitment efforts for research and development activities as well as the industry's choice for manufacturing activities.

Notes

1. See Gerald Carlino, "Declining City Productivity and the Growth of Rural Regions: A Test of Alternative Explanations," *Journal of Urban Economics* 15 (1984): 237–54.

2. John Cogan, "The Decline in Black Teenage Employment, 1950–70," *American Economic Review* 72 (September 1982): 621.

3. The Manufacturing Belt includes Massachusetts, Connecticut, New York, New Jersey, Pennsylvania, Ohio, Michigan, Indiana, and Illinois.

4. See John Rees, "Regional Industrial Shifts in the U.S. and the Internal Generation of Manufacturing in Growth Centers of the Southwest," in *Interregional Movements and Regional Growth*, ed. William Wheaton (Washington, D.C.: The Urban Institute, 1979), 51–73; and John Hekman, "The Product Cycle and New England Textiles," *Quarterly Journal of Economics* 94 (June 1980): 697–717.

5. This has been described in detail by Raymond Vernon, "Production and Distribution in the Large Metropolis," in *Locational Analysis for Manufacturing*, ed. Gerald Karaska and David Bramhall (Cambridge: MIT Press, 1969), 299–313.

6. See Edwin Mills, *Urban Economics*, 2d ed. (Glenview, Ill.: Scott Foresman, 1980), 56–77.

7. See J. N. H. Britton, "Industrial Dependence and Technological Underdevelopment," *Regional Studies* 14 (1980): 181–89.

8. Roger Schmenner, *Making Business Location Decisions* (Englewood Cliffs, N.J.: Prentice-Hall, 1982), 88.

9. E. J. Malecki, "Corporate Organization of R and D and the Location of Technological Activities," *Regional Studies* 14 (1980): 219–34.

10. See John Hekman, Mike Miles, Roger Pratt, Raymond Burby, and Anthony Marimpietri, *Impact of Environmental Regulations on Industrial Development in North Carolina* (Chapel Hill: Center for Urban and Regional Studies, University of North Carolina, 1982).

11. After removing from the study establishments that had been recorded erroneously as new or expanded, the response rate for the survey was 93 percent.

12. David Birch, *The Job Generation Process* (Cambridge, Mass.: MIT Program on Neighborhood and Regional Change, 1979).

13. *Facility Location Decisions*, a *Fortune* Market Research Survey (New York: *Fortune*, 1977).

14. See John S. Hekman, "The Future of High Technology Industry in New England: A Case Study of Computers," *New England Economic Review*, January–February 1980, 15–17.

15. North Carolina Department of Commerce, *Directory of North Carolina Manufacturing Firms, 1982–1983* (Raleigh, 1982).

16. Schmenner, *Making Business Location Decisions*, 87.

The Locational Attractiveness of the Southeast to High-Technology Manufacturers

Emil E. Malizia

Though industrial boosterism is a southern tradition that dates from the late nineteenth century, the southern states became involved more intensely in industrial recruitment and promotional activities after World War II.[1] Over the past thirty years North Carolina and other southern states appear to have experienced three "generations" of industrial development efforts. During the 1950s and 1960s when the U.S. economy was growing vigorously and mechanization of southern agriculture was displacing farmworkers, the recruitment of national manufacturers and other major industrial employers made sense as a way to offset, at least in part, the economic drain of population outmigration flowing from rural areas. The first generation of industrial development efforts offered industrial sites, financial incentives, and other forms of assistance to national companies that were seeking locations for industrial facilities. As a result, in the past fifteen years the decentralization of manufacturing has increased employment greatly in the southern states. Though evaluating the results of the first-generation strategy is difficult, southern industrial developers deserve credit for facilitating industrial development in the South. Over the years, executives of national concerns have consistently ranked the southern states among the most favorable locations for their branch plants in various opinion surveys.[2]

By the 1970s economic pressures resulting from earlier successes at industrial development began to manifest themselves. State and local leaders learned that industrial growth created, as well as solved, local problems, and the states began to articulate and embrace a second-generation industrial development strategy. Though the concern for *more* jobs continued, southern states grew more interested in attracting *better* jobs. These new efforts discriminated against certain industries and in favor of others, in seeking greater compatibility between the community's and the industry's needs. Several states, for example, opened international offices and began competing for the branch plants of particular multinational concerns. Mississippi stopped recruiting "cut-and-sew operations." South Carolina endeavored to draw incoming companies to areas of the state with relatively high rates of unemployment.

Over the past decade the national political economy has changed dramatically. National output and productivity growth has been erratic, and U.S. firms no longer dominate the international marketplace. In response, discussions of spatial policy have shifted from an emphasis on area poverty and uneven regional development to an emphasis on new technologies, innovation, and growth in productivity. Industrial policy is receiving attention while regional policy is ignored. At the state and local levels, interarea competition for corporate investments has escalated. Recruitment of industrial facilities has become a more difficult and costly endeavor.

Given these conditions, many states have begun articulating a third-generation industrial development strategy, one that aims at attracting innovative firms. Innovative firms, commercializing new products and applying new technologies, are quite volatile. Some grow rapidly while others shrink or die. Yet the survivors are believed to promise long-term economic stability and added strength to localities fortunate enough to secure them. For the United States as a whole they augur for more rapid national economic growth and for larger shares of international markets. "High-technology" firms are an important segment of such enterprises.

Numerous states and larger cities have initiated some version of an industrial development strategy directed at high-technology firms. As with the second-generation approach, these firms are supposed to offer better work, both in terms of wages or working conditions available to local workers and in terms of the positive impact on the area's industrial mix. However, with the emphasis on innovation and new product development there arises a unique concern for creating new work: new production processes and organizational arrangements must be established to produce new commodities.

The state of North Carolina is actively pursuing its own version of these industrial development strategies, introducing new approaches as older ones mature. To generate more jobs, the state continues to recruit branch plants of multilocational concerns. North Carolina appears to be one of the preferred locations for manufacturers of mature products. Since the inception of the Research Triangle Park, the state has tried to attract companies that provide better jobs. Over the past twenty years the characteristics of North Carolina's industrial mix appear to have improved. Most recently, the state has initiated a third-generation strategy emphasizing new-technology firms that are part of growing industries. In 1980 Governor Hunt articulated the policy of attracting microelectronics firms to the state. He spearheaded the establishment of the Microelectronics Center of North Carolina (MCNC) to provide a locus for basic and applied research on electronic components. Subsequently several national corporations announced plans to locate electronic facilities in or near to the Research Triangle area.

This chapter addresses three questions that are relevant to the development of North Carolina's third-generation strategy for recruiting high- or new-technology firms:

1. Which factors are uniquely important to high-technology manufacturers locating in the Southeast compared to the factors germane to all manufacturers locating in this region?
2. How do state and local efforts designed to promote (or discourage) industrial development under first- or second-generation strategies influence the location of high-technology firms?
3. What additional activities should the state consider to complement MCNC in order to attract microelectronics and other high-technology firms to North Carolina?

Research procedures and results are described below, after which follows a discussion of these three questions.

Research Procedures and Results

This research is based upon a survey of manufacturers that made capital investments in North Carolina, South Carolina, or Virginia between 1 January 1977 and 31 March 1981.[3] The manufacturing establishments were classified according to seven characteristics:

1. Location (North Carolina, South Carolina, or Virginia)
2. Type (independent firm or branch plant)
3. Size, in terms of employees (small, 0–49; medium, 50–249; large, 250 or more)
4. Industry, by two-digit Standard Industrial Code (SIC)
5. Site (rural or urban), by ZIP code
6. Growth rate (see Appendix A)
7. Level of technology (see Appendix A)

Distribution of the 136 manufacturers across these seven characteristics is presented in table 1.

The business executives who had been involved in determining the

Table 1. Manufacturing Establishment Characteristics

	Number	%
Type of Establishment		
New independent firm	30	22.1
New branch plant	106	77.9
Employment Size		
Less than 50	55	40.4
50-249	42	30.9
240 or more	39	28.7
Location		
Virginia	20	14.7
North Carolina	81	59.6
South Carolina	35	25.7
Situs		
Urban	33	24.3
Rural	103	75.7
Growth Rate		
High	42	30.9
Slow	73	53.7
Declining	21	15.4
Level of Technology		
High technology	15	11.0
Low technology	121	89.0
SIC Code--Industry		
20 Food	9	6.6
21 Tobacco	2	1.4
22 Textiles	11	8.1
23 Apparel	9	6.6
24-27 Lumber, furniture, paper and printing	16	11.8
28 Chemicals	11	8.1
29-32 Petroleum, rubber, leather, stone products	12	8.8
33-34 Primary and fabricated metals	16	11.8
35 Machinery, except electrical	19	14.0
36 Electrical machinery	16	11.8
37 Transportation equipment	8	5.9
38-39 Medical supplies and miscellaneous manufacturing	7	5.1

location of these new establishments were asked to rate the relative importance of 31 locational variables that they could have considered. These variables fall within three logical groupings that reflect the economic, spatial, and political factors usually viewed as important in making industrial location decisions: (1) access, input availability, and cost; (2) community quality of life; and (3) state or local government activities that are thought to encourage or discourage industrial development.[4] A detailed list is shown in table 2. Factor analysis, a statistical technique often used to test the validity of logical groupings and to reduce the number of variables examined in an analysis, was applied in this study to see if the 31 variables could be grouped into a smaller number of factors that would be consistent with the three logical groupings.

As shown in table 3, the results were favorable. Ten unambiguous factors summarized the 31 variables while leaving only 25 percent of the original variation unexplained. The table lists all factor loadings greater than 0.50. Notice that each variable is associated with one and only one factor; this clarifies the meaning of each factor. Factor names, shown as column headings, were inferred from these associations. Eigen values demonstrate the relative importance of each factor in capturing the variation in the original array of 31 variables. (For more detail on the factor analysis, see Appendix B.) The factors were also consistent with the three logical groupings. Two factors tapped the access and proximity variables; three covered input availability and cost variables; two combined public sector activities that influence industrial development; and three summarized the community quality of life variables. When the logical groupings of industrial location variables listed in tables 2 and 3 are compared with the groupings derived from the factor analysis in table 3, the factors reveal a high degree of consistency and appear to cover the major dimensions of the industrial location decision.

The research concentrated on contrasting the locational preferences of executives from firms that had *different* characteristics. The locational preferences of the 136 respondents were determined by their factor scores for the ten factors. The relative importance of the seven establishment characteristics was tested using discriminant analysis (see Appendix B). For example, did executives of the 30 independent firms have similar locational preferences? Were the preferences of the 106 branch plant executives similar? Were their preferences different enough to assign the branch plant executives consistently to one grouping and the independent firm executives to the other? The discriminant analysis correctly classified almost all branch plant execu-

Table 2. Industrial Location Variables

Access/Input Availability and Cost

- Labor productivity
- Skilled labor supply
- Unskilled labor supply
- Wage rates
- State/local industrial climate
- Electricity availability/cost
- Fuel availability/cost
- Transportation
- Proximity to markets
- Proximity to suppliers/business services
- Land availability/room for expansion
- Cost of land and construction

State/Local Government Support for Industrial Development

- Availability of technical training programs
- Business taxation
- State financial incentives
- Water supply
- Public wastewater treatment capacity
- Solid/hazardous waste disposal facilities
- State/local environmental regulations and permit processing

Community Quality of Life

- Housing
- Educational system
- Recreational opportunities
- Cultural resources
- Entertainment
- Personal taxes
- Cost of living
- Physical quality of air and water resources
- Aesthetic quality of natural landscape (scenery)
- Open space
- Transportation (local traffic conditions/public transportation)
- Climate

tives in one grouping but misclassified almost half of the independent firm executives. Because of that result, locational preferences based on type of establishment were not analyzed further. But the discriminant analysis did identify three important establishment characteristics: employment size, growth rate, and level of technology. The differential influence of size of firm, growth rate, and level of technology on locational preference was examined by plotting and analyzing the standardized factor scores for various executive categories.

Contrasts by size of firm are given in figure 1. Scores for executives from the 39 large establishments were much higher for land availability/cost, infrastructure availability, livability/education system, and local transportation (factors e, f, h, j), factors most relevant to executives seeking sites for large-scale facilities. These executives appear to be quite concerned about what might be called the absorptive capacity of alternative locations. The results also suggest that executives of larger establishments approached the location decision more comprehensively and systematically, as evidenced by their higher than average scores for most of the factors. The 55 executives of small establishments, on the other hand, rated all factors lower than average except access, proximity, and rural location (a, b, i). For seven of the ten factors, scores for medium-sized establishments (42 executives) were closer to those for large establishments than to those for small establishments.

Factor scores of locational preferences for establishments in nationally high-growth, slow-growth, and declining industries are compared in figure 2. Scores for the 73 slow-growth establishments are generally similar to the profiles for larger establishments, shown in figure 1. Other than the access and proximity factors, their scores are higher than the averages for all establishments. Scores for the 21 declining manufacturers are striking in that access/proximity, labor skill/productivity, and land availability/cost (factors a, c, e) are much more important than average while all other factors are much less important compared to all establishments. This may reveal the strong orientation toward efficiency in production and distribution that is expected of companies competing in nationally declining markets—for example, markets for mature products.[5] Especially surprising here is the low score for unskilled labor supply, a critical resource for the manufacture of mature products. One explanation might be that the declining industries represented in the sample were not producing such products. A more plausible one is that the executives surveyed had already decided that the Southeast has good interregional access and sufficient unskilled labor supply for the location of their facilities. This interpretation is supported by the finding that 95 percent of the executives surveyed said that they had only considered the southeastern region. It is also consistent with our understanding that industrial location is a sequential decision-making process that begins with the selection of a region or area and then proceeds to the selection of a particular com-

Table 3. Identified Factors and Factor Loading*

FACTOR GROUPINGS:

I	Access/Input Availability and Cost
II	State/Local Government Support for Industrial Development
III	Community Quality of Life

Industrial Location Variables	
Labor productivity	
Skilled labor supply	
Unskilled labor supply	
Wage rates	
State/local industrial climate	
Electricity availability/cost	I
Fuel availability/cost	
Transportation	
Proximity to markets	
Proximity to suppliers/business services	
Land availability/room for expansion	
Cost of land and construction	
Availability of technical training programs	
Business taxation	
State financial incentives	
Water supply	II
Public wastewater treatment capacity	
Solid/hazardous waste disposal facilities	
State/local environmental regulation	
Housing	
Educational system	
Recreational opportunities	
Cultural resources	
Entertainment	
Personal taxes	III
Cost of living	
Physical quality of air and water resources	
Aesthetic quality of natural landscape	
Open space	
Local traffic conditions/public transportation	
Climate	

*Only factor loadings greater than 0.50 are shown.

a. Access/proximity
b. Proximity to suppliers and business services
c. Labor skill and productivity
d. Unskilled labor supply
e. Land availability and cost
f. Infrastructure availability
g. Business taxes/financial incentives
h. Livability/educational system
i. Rural
j. Local transportation

Eigen Values									
a	b	c	d	e	f	g	h	i	j
1.6	1.1	2.3	1.4	2.4	3.3	1.2	5.5	3.1	1.2
		.79							
		.59							
			.88						
		.71							
						.57			
					.53				
					.51				
.60									
.85									
	.73								
				.79					
				.84					
							.60		
						.58			
						.76			
					.72				
					.76				
					.71				
					.73				
							.80		
							.78		
							.74		
							.82		
							.81		
							.54		
							.54		
								.59	
								.82	
								.86	
									.71
								.70	

Figure 1. Locational preferences by size of firm

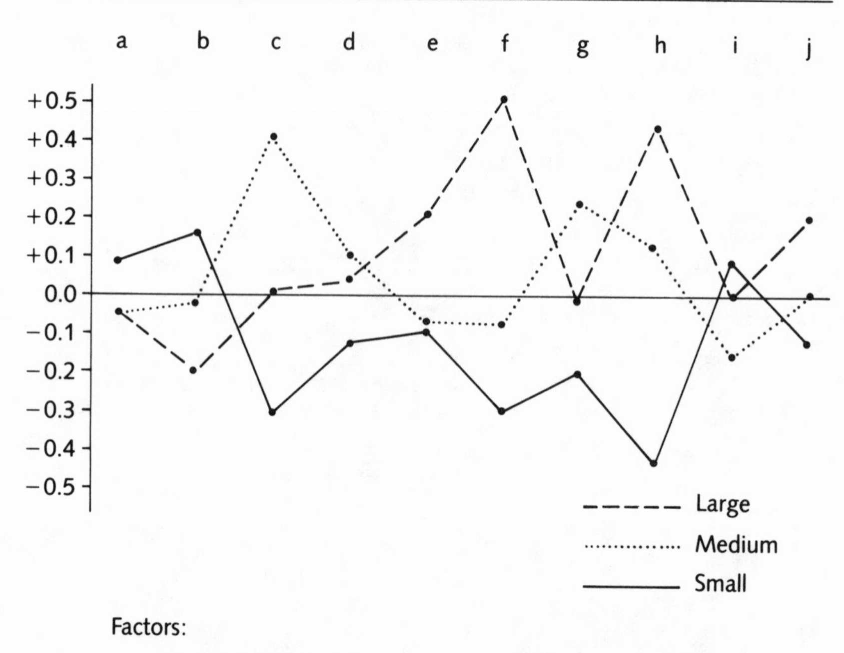

Factors:

a. Access/Proximity
b. Proximity to suppliers and business services
c. Labor skill and productivity
d. Unskilled labor supply
e. Land availability and cost
f. Infrastructure availability
g. Business taxes/financial incentives
h. Livability/educational system
i. Rural
j. Local transportation

Note: All scores are measured in standard deviation units. The horizontal line represents the average score for all 136 new establishments. The profile for each category consists of deviations above and below the mean across the ten categories.

munity or site.[6] Scores for the 42 high-growth industries are intriguing on several counts. First, proximity to suppliers and business services is the only score significantly above the all-establishment average. Second, the profile resembles that for small establishments, shown in figure 1. But it is unclear whether the executive preferences shown in these profiles are more germane to executives starting a business in a high-growth industry, to those starting a small business, or to some other, more basic, characteristics.

Figure 2. Locational preferences by growth rate

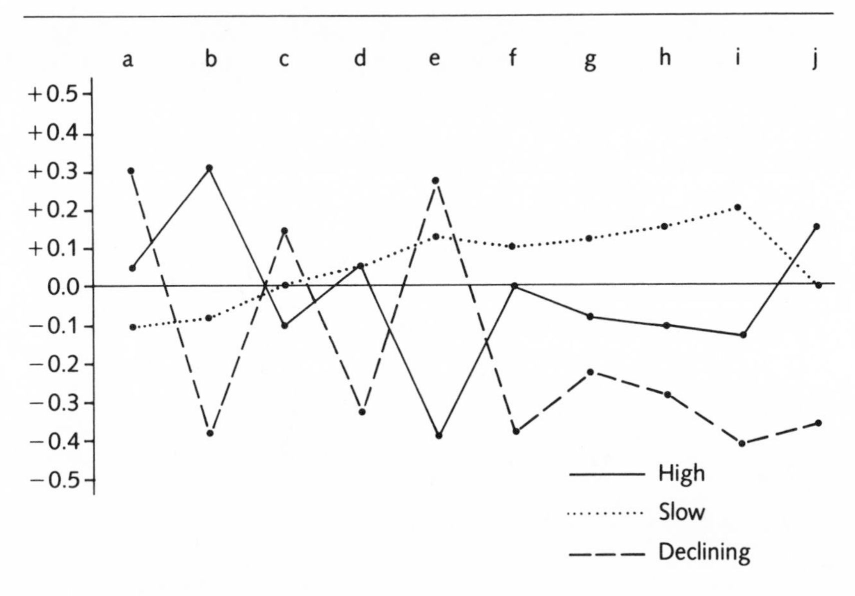

Factors:

a. Access/Proximity
b. Proximity to suppliers and business services
c. Labor skill and productivity
d. Unskilled labor supply
e. Land availability and cost
f. Infrastructure availability
g. Business taxes/financial incentives
h. Livability/educational system
i. Rural
j. Local transportation

Note: All scores are measured in standard deviation units. The horizontal line represents the average score for all 136 new establishments. The profile for each category consists of deviations above and below the mean across the ten categories.

High-technology and low-technology establishments are contrasted in figure 3. Because 121 of the 136 establishments were assigned to the low-technology category, their profile is not very different from the all-establishment averages. Executives from these establishments appear to view availability of unskilled labor, land availability/cost, and business taxes/financial incentives (factors d, e, g) as more important than average, and infrastructure availability, livability/education system, and local transportation (factors f, h, j) as less important. Executives from the 15 high-technology establishments revealed very differ-

Figure 3. Locational preferences by level of technology

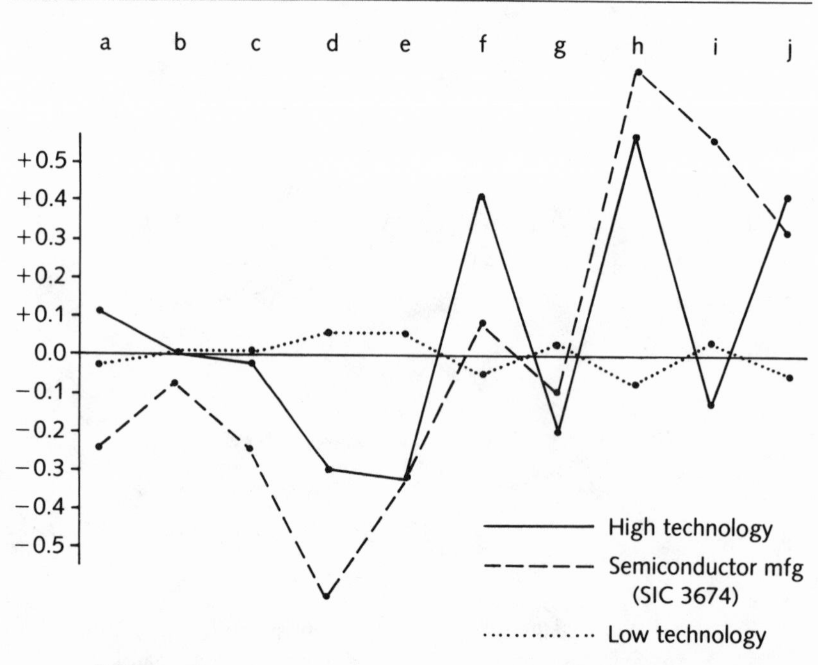

Factors:

a. Access/Proximity
b. Proximity to suppliers and business services
c. Labor skill and productivity
d. Unskilled labor supply
e. Land availability and cost
f. Infrastructure availability
g. Business taxes/financial incentives
h. Livability/educational system
i. Rural
j. Local transportation

Note: All scores are measured in standard deviation units. The horizontal line represents the average score for all 136 new establishments. The profile for each category consists of deviations above and below the mean across the ten categories.

ent preferences. For them the livability/education system factor is most important, followed by local transportation and infrastructure availability. Land availability/cost, unskilled labor supply, and business taxes/financial incentives are least important. In addition, these high-technology executives rated access/proximity (factor a) more important than average and rural orientation (factor i) less important. The results confirm the general impression that high-technology concerns have locational preferences substantially different from other concerns. Their executives place particular emphasis on the community quality of life variables (factor h) when making location decisions (see table 3).

Because none of the establishments in the sample represented the microelectronics industry (specifically SIC 3674, manufacturers of semiconductors and related devices), several industry experts were consulted to learn about the locational preferences of such firms. The best information was provided by the Semiconductor Industry Association of Cupertino, California. The executive director was gracious enough to respond to the same survey questions that generated the information analyzed in this research. On the basis of his responses, a profile of a "typical" semiconductor manufacturer has been supplied in figure 3 along with those for high-technology and low-technology establishments. The most striking feature of this profile is that the community quality of life factors (h, i, j) would appear to be most important to executives locating semiconductor manufacturing facilities. The livability/educational system is particularly important. All other factors, except infrastructure availability (factor f), assume negative values, indicating their relative unimportance. When compared with others in figure 3, the semiconductor manufacturer's profile appears generally to follow the high-technology profile, except for the access/proximity and rural factors (a and i). Otherwise, the locational preferences of executives of high-technology establishments located in the Southeast appear quite similar to those we assumed for semiconductor industry executives.

The distribution of high-technology establishments according to size of firm and growth rate is interesting. There are 3 small, 5 medium-sized, and 7 large establishments in the sample. None is in declining industries: 11 of the 15 are in high-growth industries, the other 4 in slow-growth industries. Unfortunately the sample is too small to permit further useful analysis of such subsamples.

Policy Issues

The three policy questions posed above can be addressed using the profiles shown in figures 1–3. Results clearly show that executives of high-technology manufacturers are interested in a particular group of regional location factors different from those preferred by executives of other manufacturing facilities. A community's livability, education system, local transportation, and infrastructure are most important to this group of executives, as shown in figure 3. These results are partly consistent with the Joint Economic Committee's recent findings that skilled labor, moderate taxes, and universities are the features that attract high-technology companies to a region.[7]

State and local activities designed to influence industrial development seem to have varying importance in attracting high-technology establishments. The six locational variables that dominate the infrastructure availability factor (see table 3) appear to be important to executives locating high-technology establishments in the Southeast; business taxes and financial incentives are relatively unimportant. These results are not consistent with the Joint Economic Committee's study, which emphasizes the importance of taxation and deemphasizes the importance of infrastructure.[8] A safe conclusion is that some, but not all, state and local activities developed under the first- and second-generation industrial development strategies continue to be influential in attracting high-technology firms. This result also seems to hold for semiconductor manufacturers (compare factors f and g in figure 3).

The locational profiles also suggest some goals for new third-generation industrial development strategies. First, the state should place emphasis on community quality of life factors. Indeed, to attract microelectronics firms, the local quality of life must be sufficiently high to satisfy the scientists, engineers, and technical personnel needed to sustain the growth of a microelectronics sector in North Carolina. Particularly important is the quality of both graduate education and vocational education, which together offer advanced educational opportunities for all segments of the industry's work force. The Joint Economic Committee's study likewise emphasizes the importance of universities and educational facilities as an attraction to high-technology firms.[9]

Tax and financial incentives appear to be less important in general than other factors. However, new financial programs may be necessary to effectively foster the growth of high-technology industries in the state. The financing offered through North Carolina state agencies can play important roles in third-generation industrial development efforts.

Local governments should focus on improving primary and secondary education, providing affordable housing, and developing local amenities in a fiscally responsible manner. Good planning and growth management should be viewed not only as the means to mitigate adverse effects of industrial growth, but also as essential tools for enhancing community quality of life to help attract microelectronics to the state in the first place. Thus, by continuing some traditional activities and emphasizing new ones, the state and local governments in North Carolina should be able to improve their chances of attracting and successfully accommodating high-technology manufacturers.

Appendix A

The 1981 *U.S. Industrial Outlook* published by the U.S. Department of Commerce, Industry and Trade Administration, provides information on the real value of shipments from 1972 through 1978 for each manufacturing sector. These data were used to compute real annual growth rates for each four-digit SIC industry in the sample. From the resulting frequency distribution of industry growth rates, the establishments were assigned to one of three annual growth rate categories: high (above 5.0 percent), slow (between -0.5 percent and 5.0 percent), or declining (below -0.5 percent).

Textual information from recent issues of *U.S. Industrial Outlook* and national-level statistics for 1967–77 published by the U.S. National Science Foundation, *National Patterns of Science and Technology Resources* (annual), were used to determine which establishments in the sample would qualify as being in high-technology industries. The key referents were expert opinion on innovation and technological sophistication of the industry, total research and development expenditures and basic research expenditures as a percentage of net sales, and scientific, engineering, and technical personnel per 1,000 employees.

Numerous references on technological change, innovation, and the commercialization of inventions were consulted, as well as several descriptions of state government innovation strategies, to arrive at a proper definition of high-technology industries. Not surprisingly, the term is neither defined nor used precisely. The popular image is that these industries are attractive because they provide good jobs but do not pollute the environment, use much energy, require expensive infrastructure, or have collective bargaining agreements. Clearly, research and development efforts influence the growth of technology-based products in very complex ways. Scientific discoveries in electronics and genetics, for example, have given rise to new products and processes. As such innovation creates new industries, the industrial classification system that was designed to codify industrial structure itself must be changed. Differences within specific industries also exist. For example, company-level information on research and development activities is available in *Business Week*, 5 July 1982.

In this research, ten industries represented in the survey were considered to be nationally high-technology industries: industrial gases (2813); medicinal chemicals and botanical products (2833); pharmaceutical preparations (2834); industrial organic chemicals not classified elsewhere (2869); pesticides and agricultural chemicals not classified elsewhere (2879); special dies, tools, and industrial molds (3544); electronic computing equipment (3573); telephone and telegraph apparatus (3661); radio and television transmitting, signaling, and detection equipment and apparatus (3662); and electronic components not classified elsewhere (3679).

Appendix B

Factor analysis was used to summarize the locational variables; results are shown in table 3. As noted, ten factors explained 75 percent of the total variation among the 31 variables. A varimax rotation was performed to clarify the factor loadings. The rotation generated seven group factors and three specific factors with eigen values greater than 1.0. Only factor loadings greater than 0.50 are shown in the table.

Discriminant analysis was applied to each of the seven establishment characteristics to determine whether or not categories (such as small, medium, large size) within each would consistently reveal meaningful differences among locational preferences. Meaningful categories for industry by SIC could not be established. The discriminant analyses successfully classified establishments for three of the six other characteristics. Locational preferences examined in terms of categories of size, growth rate, and level of technology turned out to be significantly different. The proportion of all establishments correctly classified was consistently high for size (75 percent), growth rate (81 percent), and level of technology (98 percent). Analysis by location (by state) and site (urban or rural) did not successfully define distinguishable categories. Results for establishment type (branch plant or independent firm) were much more significant than those for location or site but less significant than those for the other three characteristics; because only 53 percent of independent firms were assigned to the correct category, this characteristic (type) was excluded from further analysis.

Influence of establishment type had been analyzed in two previous studies: Emil E. Malizia, "Making Communities Attractive to Business or Improving the Quality of Life," in *Urbanization and Technology*, Proceedings of the Fourth Annual Urban Affairs Conference, ed. W. J. Wicker (Chapel Hill: University of North Carolina, Urban Studies Council, 1982), 12–26, and "Contrasts in the Locational Attractiveness of the Southeast to New Manufacturers," *Review of Regional Studies* 12 (1983). Executives locating branch plants were found to have preferences different from those locating independent firms. In the results reported in this chapter, the factor score profile for branch plants resembled the profiles for large establishments and establishments in slow-growth and low-technology industries. The profile for independent firms resembled those for small establishments and establishments in high-growth or high-technology industries.

The influence of the three discriminating characteristics (size, growth rate, level of technology) on the ten locational factors was further examined using a general linear model that permitted the inclusion of all possible interaction effects. The model took the following form:

$$(F1, F2, \ldots, F10) = f(E, G, T, E^{*}G, E^{*}T, G^{*}T, E^{*}G^{*}T),$$

where $F1$–$F10$ = rotated factors; E = employment size; G = growth rate; and T = level of technology. The seven independent effects specified in the general linear model were tested using multiple analysis of variance (MANOVA) procedures. Employment size turned out to be the most significant effect, beyond the 1 percent level; growth rate was significant beyond the 2 percent level; level of technology was significant beyond the 9 percent level. On the other hand, none of the four combined independent effects yielded statistically significant results. The combination that most closely approached significance was that of size, growth rate, and level of technology, at 13 percent. On the basis of these results the factor scores were analyzed only for the separate independent effects—size, growth rate, and level of technology—without regard to the four interaction terms.

Each of the three independent effects found significant in the discriminant analysis was analyzed by plotting the factor scores for each category within size, growth rate, and technology. Each category appeared to have a distinctive profile of locational preferences that made sense given the interpretation of the underlying ten factors.

For further discussion of the research methods, see Malizia, "Contrasts in the Locational Attractiveness of the Southeast" (above).

Notes

1. See J. C. Cobb, *The Selling of the South: The Southern Crusade for Industrial Development, 1936–1980* (Baton Rouge: Louisiana State University Press, 1982).

2. See, for example, Alexander Grant & Co., *A Study of Manufacturing Business Climates in the 48 Contiguous States of the U.S.* (Washington, D.C., 1980).

3. The survey was conducted by the Center for Urban and Regional Studies and the School of Business, University of North Carolina at Chapel Hill, for

the North Carolina Department of Natural Resources and Community Development.

4. See D. Z. Czamanski, "A Contribution to the Study of Industrial Location Decisions," *Environment and Planning A* 13 (1981): 29–42; B. M. Moriarty, *Industrial Location and Community Development* (Chapel Hill: University of North Carolina Press, 1980), 89–145; and R. W. Schmenner, *Making Business Location Decisions* (Englewood Cliffs, N.J.: Prentice-Hall, 1982).

5. See N. Hansen, "The New International Division of Labor and Manufacturing Decentralization in the United States," *Review of Regional Studies* 9 (1979): 1–11; and J. S. Hekman, "The Product Cycle and New England Textiles," *Quarterly Journal of Economics* 94 (1980): 697–717.

6. See Moriarty, *Industrial Location and Community Development*, 89–100.

7. U.S. Congress, Joint Economic Committee, *Location of High Technology Firms and Regional Economic Development* (Washington: Government Printing Office, 1982), 49–56.

8. *Location of High Technology Firms*, 22–28.

9. *Location of High Technology Firms*, 49–51.

Part Three

Planning for the Microelectronics Industry in North Carolina: The State Perspective

8 The States and High-Technology Development: The Case of North Carolina

Michael I. Luger

Much has been written recently about North Carolina's bold high-technology development initiatives; its Research Triangle Park, School for Science and Mathematics, and Microelectronics Center are now considered standards to be copied.[1] More recently the state has renewed its commitment to establish a Biotechnology Center and has also begun to address the needs of small high-technology businesses. These efforts contribute to North Carolina's image as a pioneer in high-technology development. However, these few visible programs do not by themselves constitute a sound economic development strategy. It is important to place North Carolina's high-technology initiatives within a broader policy perspective. A review of the full range of economic development programs used in North Carolina shows that a disproportionate amount of attention has been paid to high-technology recruiting in comparison with retention and creation of resident high-technology and traditional businesses. Comparing North Carolina's initiatives with those of other states illustrates that North Carolina lags in the use of some cost-effective and innovative programs.

Policy Goals

State documents, disseminated during the tenure of Governor James B. Hunt, Jr. (1976–84), identified four economic development goals for North Carolina: higher wages, more jobs, greater job stability, and more balanced geographic outcomes. In addition, Governor Hunt himself specified that any development strategy must be fiscally sound.[2]

The author has benefited from comments on an earlier draft by Wim Wiewel, Tom Rudin, Quentin Lindsey, and Dale Whittington, and from membership on the Long Term Resources Committee of Governor Hunt's Task Force on Science and Technology. The views expressed in the paper are the author's alone and should not be attributed to the Task Force.

Hunt's successor, Republican Jim Martin, has stressed those same goals.[3]

Higher wages are particularly important, because North Carolina has the lowest average hourly manufacturing wage rate in the country.[4] Because the state also has one of the largest manufacturing sectors, those low manufacturing wage rates have added significance. Not surprisingly, North Carolina's per capita income is only 85 percent of the national average.[5]

More jobs are critical in North Carolina for three reasons. First, the state has one of the fastest-growing populations in the country (almost 2 percent per year over the past decade); unless job opportunities keep pace with the influx of workers, unemployment will increase. Second, as existing businesses modernize they are apt to become more capital-intensive, so new jobs must be found for workers who are displaced or denied jobs in those businesses. Third, the state has relatively large numbers of new small enterprises and old branch plants in declining industries—the two types of businesses most likely to fail—and its rate of plant closings is high.[6] From 1978 through 1982, for example, 287 manufacturing plants closed permanently in North Carolina, eliminating jobs for 27,131 employees. Another 244 plants closed temporarily, putting more than 36,000 workers on the short-term unemployment rolls.[7] New jobs are needed to absorb those workers.

Greater job stability is important because North Carolina has been more vulnerable than the nation as a whole to most postwar business cycles, in large part because its economic base is dominated by a few procyclical industries.

Balanced development is important for North Carolina because wage and unemployment rates vary considerably among its counties (see figure 1). This problem was recognized by the General Assembly in 1979 when, at Governor Hunt's request, it passed the North Carolina Balanced Growth Policy Act, which committed the state "to encourage diversified job growth in different areas . . . , with particular attention to those groups which have suffered from high rates of unemployment or underemployment, so that sufficient work opportunities at high wage levels can exist where people live."[8]

The last objective for development policy—that it be *fiscally sound*—is a response to recent economic events; it also reflects North Carolina's traditional fiscal conservatism. Like most states, North Carolina is bound by law to maintain a balanced budget. That has been particularly difficult for most of the past ten years, for two reasons. First, from 1975 to 1982 economic conditions generally were unfavorable for state governments. During the recessions of 1975–76 and 1978–81 tax revenues typically fell, but the demand for state-supported welfare

Figure 1. Geographic Dispersal of Wage and Unemployment Rates in North Carolina, 1978 and 1980

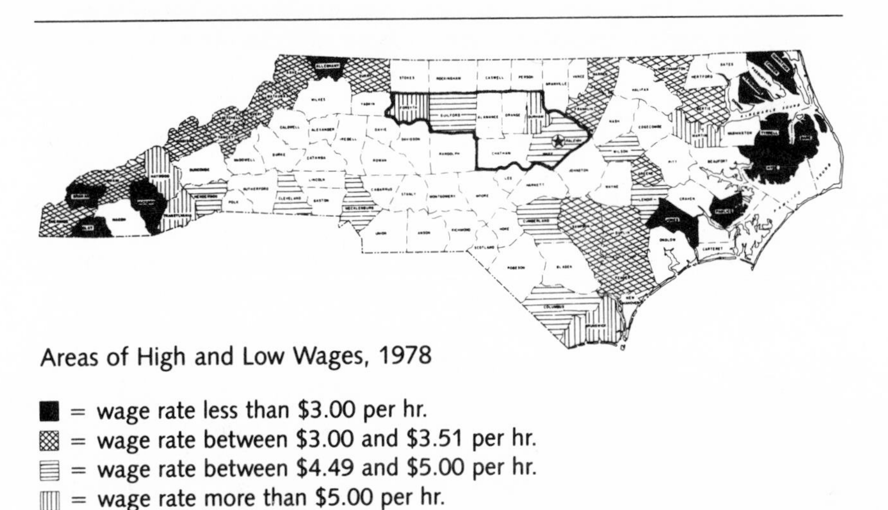

Areas of High and Low Wages, 1978

■ = wage rate less than $3.00 per hr.
▩ = wage rate between $3.00 and $3.51 per hr.
▤ = wage rate between $4.49 and $5.00 per hr.
▥ = wage rate more than $5.00 per hr.

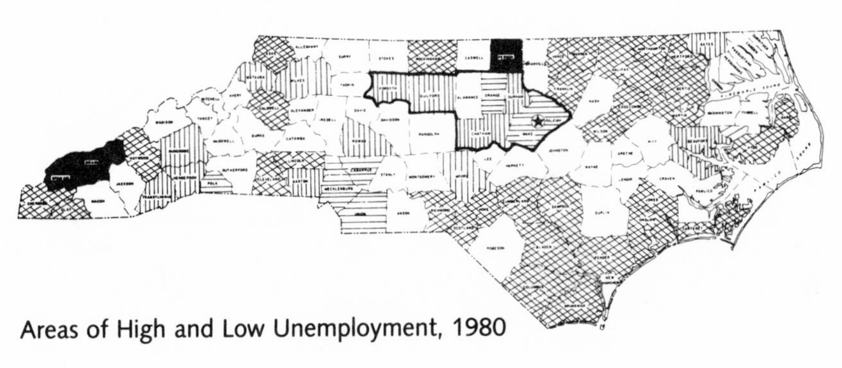

Areas of High and Low Unemployment, 1980

■ = unemployment rate over 10%
▩ = unemployment rate between 8% and 10%
▤ = unemployment rate between 5% and 6%
▥ = unemployment rate less than 5%

In the maps shown above, the heavy dark line around a seven-county area stretching from Wake to Guilford surrounds the "projected location zone" for new microelectronics facilities.

Source: Reprinted by permission from *North Carolina Insight*, magazine of the North Carolina Center for Public Policy Research.

programs, including food stamps, unemployment compensation, aid to families with dependent children, and medicaid, tended to rise. Inflation made matters worse, because per capita payments under those programs are indexed. Second, since 1981 much responsibility for economic development has shifted from Washington to state capitals, under President Reagan's New Federalism. There has not been a concomitant increase in federal subventions. Transfers from Washington to state and local governments have actually fallen: in 1981 they amounted to 3.3 percent of GNP; they are projected to be only 1.8 percent of GNP in 1985.[9] In addition, the 1981 and 1982 tax acts reduced many states' own tax receipts. By one estimate, the forty-five states, including North Carolina, that use the federal government's definition of income when calculating state tax liability will lose $19 billion in revenue between 1983 and 1987 as a consequence of liberalized depreciation guidelines in that federal tax legislation.[10] A report by the North Carolina Center for Public Policy Research has documented the particular impact of those changes in North Carolina. In fiscal 1982—the first year of the Reagan presidency—North Carolina received over $241 million less in federal aid than it did in fiscal 1981.[11] And between 1983 and 1987 the state is expected to collect $450.5 million less in tax revenues than it would have if federal tax laws had not been changed, or if the state did not use the federal government's definition of income.[12]

The State's Response

North Carolina's policy-makers have addressed these five development goals by enacting programs designed to induce existing businesses to relocate or to expand operations within the state, foster innovation in the state's traditional industries, and facilitate new business development.[13] These are of two types: (1) financial programs that provide direct loans, subsidies, and grants ("direct financial assistance"), and tax credits, deductions, and exemptions ("tax-related financial assistance"); and (2) nonfinancial programs, not directly pecuniary, such as job training and technical assistance, and programs that help create a favorable business climate. Most programs in either category can be applied generally or targeted according to the size, organizational structure, industrial classification, or location of business.

Recruiting Nonresident Businesses and Helping Existing Enterprises Grow

North Carolina has a long history of recruiting outside industries to establish branch plants in the state.[14] Governor Hunt is estimated to have spent 25 percent of his time courting businesses.[15] Other agencies and organizations have also played a role. For instance, the North Carolina Department of Commerce employed twenty-eight full-time and several part-time recruiters in 1982, and quasi-public and private development organizations—including hundreds of local chambers of commerce, banks, railroad companies, and investor-owned utilities—have also developed recruiting staffs.[16] These recruiting efforts have been targeted at foreign-owned as well as domestic businesses. In 1980 North Carolina spent nearly one-half million dollars on reverse investment promotion, which placed it eighth among all states in per capita terms.[17]

The commitment to recruit is backed up by a sizable list of inducements. Most important, the state promotes itself as having a good business environment. One recent advertisement in a national journal says: "Absenteeism and tardiness rates are among the lowest in the United States, and no state ranks better in terms of time lost due to work stoppages. A right-to-work law state, North Carolina's rate of unionization is the lowest in the nation. North Carolina's workers compensation insurance rates are also among the lowest in the nation." In addition, manufacturing wages in North Carolina are among the least costly in the country, and the state's cultural amenities are rapidly being improved. For these reasons, several national surveys have ranked North Carolina as one of the most desirable places in the United States to conduct business.[18]

Besides providing a good business climate, the state offers financial incentives. It allows local governments to issue industrial revenue bonds (IRBs), almost $1.5 billion of which were issued from 1976 to 1982.[19] It also arranges loans for businesses' general use through the Small Business Administration (SBA, Section 503) and the Farmers Home Administration, and for businesses' use in constructing new buildings through a $500,000 Speculative Building Revolving Loan Fund. In 1982–83, SBA loans to North Carolina businesses totaled almost $4 million, and between 1961 and 1981, forty-two speculative buildings were completed for businesses employing approximately 7,000 workers.[20] State officials also have used Urban Development Action Grants to help develop industrial property, often in conjunction with locally issued IRBs.[21] North Carolina exempts several types of goods, fuels, materials, and equipment from state sales and local property taxes and allows an income tax credit for part of the inventory tax.[22] Governor Martin has made the elimination of the inventory tax one of his top priorities for 1985. In addition to these direct financial and tax-related assistance programs, the state helps industry by providing community college students with specialized job training that is tailored to businesses' specifications. Almost $2.3 million was spent for that purpose in fiscal 1982.[23] The programs just described are available to all recruited businesses. They also provide an incentive for existing firms to expand their facilities in North Carolina rather than in other states.

North Carolina has developed additional programs specifically for high-technology businesses. This targeting goes back to the mid-1950s, when the Research Triangle Park was first conceived as a site for technologically advanced "clean" industries. Recruitment since then has brought more than forty major projects to the park, including branch plants of IBM, General Electric, Data General, and Troxler Electronics. This commitment to focus recruitment efforts on high-technology businesses was affirmed and strengthened under Governor Hunt. In a 1980 paper on economic development, and again in his 1983 State of the State address, he declared his intention to "continue North Carolina's leadership in attracting the high technology industries that will create the greatest number of jobs: electronics, computers, and information-processing."[24] A 1983 report by the Governor's Task Force on Science and Technology recommended that an Assistant Secretary of Commerce for New Technology be appointed "to lead recruitment efforts targeted at new technology firms and headquarters."[25]

An early strategy to induce such firms to move to the Research Triangle Park was to establish public and quasi-public "anchors" there. Whittington (chapter 1, this volume) notes that former governors Luther Hodges and Terry Sanford used their political influence to lure a U.S. Forest Service facility, the National Center for Health Statistics Laboratory, the National Institute of Environmental Health Sciences, and the U.S. Environmental Protection Agency to the park between 1962 and 1968. Those installations helped legitimize the park as a prime site for research and development in science and technology. The latest strategy to induce high-technology firms to move to the park and nearby areas has been to establish state-supported research centers as magnets. The Microelectronics Center of North Carolina was approved by the legislature in the spring of 1981, and a Biotechnology Center was established through the state's Board of Science and Technology in November 1981, similarly designed to "serve as the focal point of the state's efforts to be at the forefront of technological developments" by providing a physical location for private firms to develop new products cheaply, drawing on the resources of nearby universities.[26] Both centers also symbolize the state's commitment to a favorable high-technology business environment.

Another recent strategy for attracting high-technology companies to the Research Triangle and elsewhere around the state has been North Carolina's commitment to improving technical education in high schools, community colleges, and universities. Governor Hunt included an additional $1 million in the 1983–85 state budget to employ 350 high school science and math teachers for an additional six weeks a year, and almost $8 million for new equipment, including microcomputers. At the community college level, the state instituted special high-technology training programs at a cost of more than $3 million during the 1983–85 biennium, and dedicated $12.4 million to the purchase of new scientific and technical equipment. Governor Martin has recommended similar expenditures for the 1985–87 period. At the university level, the state is providing money for graduate fellowships and additional faculty pay for engineers and computer scientists through the Microelectronics Center and is expanding its outlays for new engineering facilities.[27] This large-scale improvement of technical education is a form of nonfinancial business assistance. But it also has symbolic importance, thus contributing to the perceived business environment. The state-supported School for Science and Mathematics for gifted high school students, which opened in Durham in 1981, is a highly visible case in point.

Making Existing Businesses Technologically More Advanced

Despite the influx of new-technology firms over the past several decades, most jobs in North Carolina are still in industries using traditional technologies, most notably textiles, apparel, and furniture. Those businesses alone provided almost 50 percent of all manufacturing jobs and over 20 percent of total jobs in June 1983.[28] Tobacco-related businesses, which also use traditional processes, account for another large proportion of jobs. In sum, less than 7 percent of all jobs in North Carolina are in high-technology businesses; approximately 30 percent are in what have been described as "low technology."[29]

As is now well known, traditional industries have become less competitive in world markets and have experienced considerable job loss over the past several decades. Between 1978 and 1982 there were more than 13,000 permanent layoffs in North Carolina's three leading traditional industries.[30] In 1982 alone, 17,000 textile workers were idled, either temporarily or permanently.[31]

The state has done little to help stabilize its declining industries. Increases in competitiveness that have occurred have resulted, in large part, from private sector initiatives.[32] Governor Hunt's Task Force on Science and Technology recommended a number of programs for legislative consideration, including the creation of a Technological Development Fund to support industry studies of the state of technology, as well as new research programs in agriculture, but the fate of these is uncertain.[33] Governor Martin recently appointed an Assistant Secretary of Commerce for Traditional Industries to focus attention on the state's declining businesses, but it is too early to assess the success of that new office.

Preserving Existing Small Businesses and Creating New High-Technology Enterprises

Small businesses (with fewer than 100 employees), both traditional and high-technology, make up 97 percent of the business population in North Carolina.[34] Their employment share is also high—as much as 70 percent by this author's estimate.[35] Small businesses are significant, however, for other reasons as well. For example, small, single-plant firms are usually locally owned, and if they can survive economically, they are less likely to be moved long distances than are branch plants of larger firms.[36] But small businesses may be more vulnerable to business cycles in comparison with their big business counterparts.[37] It is thus unclear whether new and small businesses offer more, or less, job stability than larger, more established companies. Nonetheless, small businesses are essential for a healthy state economy, for they provide jobs for workers displaced from failing businesses, often serve as vehicles for the transmission of new technologies, and heighten competitiveness.

A wide variety of general-purpose financial and nonfinancial programs benefit both small and large businesses. In addition, several technical assistance programs for small businesses are now in place: the Department of Community Colleges operates fourteen Small Business Centers around the state; there is a new Small Business and Technological Development Center within the UNC system; and the UNC Business School offers extension courses for owners and managers of small businesses. The state is less active in the provision of small business capital. Two capital programs were enacted as part of the 1983 New Technology Jobs Bill, but they are small in size, amounting to only $875,000 for the 1983–85 biennium. The first of these programs is the North Carolina Innovation Fund, a pool of money to be used as equity financing for research projects conducted by new and existing small businesses in the state. The second is an Incubator Facility Program, under whose jurisdiction the North Carolina Technological Development Authority—a new body with a twelve-member board of directors, seated within the state Department of Commerce—makes grants to help local communities build structures to house small businesses during their early phases of development. Under this plan the local developer—required by law to be a nonprofit organization—must provide secretarial and administrative help and technical assistance to the tenant businesses. Governor Hunt's Task Force on Science and Technology also recommended the creation of an Innovation Loan Program that would establish a self-sustaining revolving fund from which low-interest loans could be made to companies with innovative new products. The fate of this in the legislature is also uncertain.

Comparing States' Development Strategies

North Carolina's development strategy, outlined above, can now be reviewed from a broader policy perspective of the major development efforts of all fifty states. The nationwide inventory of state programs presented in table 1 is a compilation of data from several sources. The table lists sample financial and nonfinancial programs and indicates whether states target geographic areas, small businesses, or particular industries (including high technology). The information in the table cannot be used to compare the *effectiveness* of state development strategies, but from it one can make some judgments about the breadth of state "effort."[39]

The Most Popular Programs

The most popular sets of programs are locally issued industrial revenue and general obligation bonds (IRBs and GOBs), state and local tax abatements and exemptions, and manpower training.

Forty-five states allow municipalities to issue IRBs or GOBs; the five other states issue bonds on behalf of their municipalities (see columns in table 1 for IRBs, GOBs, and umbrella bonds).[40] The rules governing the issuance of bonds differ from state to state. In many cases bonds can be used for virtually any business project, but in several states

Table 1. State Development Programs

	FINANCIAL PROGRAMS									
	Direct Assistance									
	Land & Bldg Subsidies	Direct Grants	Direct Loans	Private Development Credit Corporation	State Funded/Chartered Equity/Venture Co.	Loan Guarantees	Interest Rate Subsidies	State Issued Ind. Rev. or Gen. Obligation Bonds	Locally Issued Ind. Rev. or Gen. Obligation Bonds	State or Locally Issued Umbrella Bonds
Alabama	x	x							x	
Alaska			x		x					x
Arizona									x	
Arkansas				x		x			x	
California			x			x		x	x	
Colorado									x	
Connecticut			x		x	x		x		x
Delaware								x	x	
Florida			x	x					x	
Georgia				x					x	
Hawaii			x					x		
Idaho									x	
Illinois			x					x	x	
Indiana	x	x	x		x	x			x	
Iowa				x					x	x
Kansas				x					x	
Kentucky			x	x		x		x	x	
Louisiana			x		x	x			x	
Maine					x	x		x	x	x
Maryland				x		x		x	x	x
Massachusetts				x	x	x		x	x	
Michigan				x	x			x	x	x
Minnesota			x			x			x	x
Mississippi						x			x	
Missouri	x			x		x	x	x	x	x

FINANCIAL PROGRAMS									NONFINANCIAL PROGRAMS					TARGETING		
Tax-Related Assistance																
No Sales/Use Tax	Investment Tax Credit	Jobs Tax Credit	R&D Tax Credit or Exemptn	Property Tax Abatement	Bus. Inventories Exemptn	Goods-in-Transit Exemptn	Fuels/Materials Exemptn	Industrial Equip. Exemptn	Employee Job Training	Procurement Assistance	Development Task Force	Research Park	Right-to-Work Law	Geographic, including Enterprise Zones	Small Businesses	Favored Industries (o = high technology)
				x	x	x	x	x	x					x		
x	x				x											x
					x	x		x				x	x			x
				x		x	x	x	x				x			x
		x	x		x	x	x	x	x			x		x	x	o
	x	x		x			x	x	x						x	o
		x		x	x	x		x	x			x		x	x	o
	x	x			x	x	x	x	x			x		x		
	x	x		x	x	x	x	x	x			x	x	x		o
					x	x	x	x	x				x			o
				x		x	x	x	x						x	o
	x	x			x	x	x	x	x							x
	x			x		x	x	x	x			x		x		o
	x	x		x		x	x	x	x							o
		x		x	x	x	x	x	x				x		x	x
	x	x			x	x	x	x	x				x	x		
						x	x	x	x					x	x	x
				x		x			x				x	x	x	x
	x	x		x	x	x	x	x							x	x
		x		x	x	x	x	x	x			x		x	x	x
	x	x		x	x	x	x	x	x	x		x		x	x	o
	x			x	x	x	x	x	x						x	o
				x	x	x	x	x		x				x	x	x
				x	x	x	x	x	x			x	x		x	o
	x	x		x	x	x	x	x	x		x			x	x	o

Table 1, continued

	FINANCIAL PROGRAMS									
	Direct Assistance									
	Land & Bldg Subsidies	Direct Grants	Direct Loans	Private Development Credit Corporation	State Funded/Chartered Equity/Venture Co.	Loan Guarantees	Interest Rate Subsidies	State Issued Ind. Rev. or Gen. Obligation Bonds	Locally Issued Ind. Rev. or Gen. Obligation Bonds	State or Locally Issued Umbrella Bonds
Montana			x	x				x	x	
Nebraska				x				x	x	x
Nevada								x	x	
New Hampshire				x		x		x	x	
New Jersey		x	x			x		x		
New Mexico									x	
New York			x	x	x	x		x	x	x
North Carolina	x				x				x	
North Dakota	x			x		x			x	
Ohio		x	x			x		x	x	
Oklahoma			x						x	
Oregon			x					x	x	
Pennsylvania			x						x	
Rhode Island						x		x		
South Carolina				x					x	
South Dakota									x	
Tennessee									x	
Texas			x						x	
Utah									x	
Vermont			x			x		x	x	
Virginia									x	
Washington									x	
West Virginia			x					x	x	
Wisconsin					x				x	
Wyoming									x	

FINANCIAL PROGRAMS									NONFINANCIAL PROGRAMS					TARGETING		
Tax-Related Assistance																
No Sales/Use Tax	Investment Tax Credit	Jobs Tax Credit	R&D Tax Credit or Exemptn	Property Tax Abatement	Bus. Inventories Exemptn	Goods-in-Transit Exemptn	Fuels/Materials Exemptn	Industrial Equip. Exemptn	Employee Job Training	Procurement Assistance	Development Task Force	Research Park	Right-to-Work Law	Geographic, including Enterprise Zones	Small Businesses	Favored Industries (o = high technology)
x	x	x		x	x							x			x	x
					x	x	x	x	x				x		x	x
					x	x	x						x			
x				x	x			x								x
				x	x	x	x	x	x		x	x		x		x
					x	x	x	x	x							o
	x	x		x			x	x	x					x	x	o
					x	x	x	x	x		x	x	x		x	o
	x	x		x	x	x	x		x				x			x
	x	x		x		x	x	x	x	x				x	x	o
	x	x		x		x	x	x	x							x
x					x		x					x				x
	x			x	x		x	x	x		x			x	x	o
				x	x	x	x	x	x							o
				x	x	x	x	x	x				x			x
				x	x	x	x	x	x				x			x
	x			x	x	x	x	x	x			x	x			o
				x		x		x	x			x	x			o
					x	x	x	x				x	x			
				x	x		x	x	x							x
	x	x		x		x	x	x	x		x		x	x		o
					x	x	x	x				x				o
	x						x	x	x							x
	x					x	x	x						x	x	
					x	x	x	x					x			

Sources: "15th Annual Report of Legislative Climates," Industrial Development, March-April, 1981; "16th Annual Report of Legislative Climates," idem, January-February 1982; U.S. Office of Technology Assessment, Technological Innovation and Regional Economic Development, Background Paper 1 (Washington, D.C., 1983); U.S. Office of Technology Assessment, Technology, Innovation and Regional Economic Development: Encouraging High Technology Development, Background Paper 2 (Washington, D.C., 1984); National Governor's Association, Technology and Growth (Washington, D.C., October 1983); Urban Institute et al., Directory of Incentives for Business Investment and Development (Washington, D.C., 1983); Michael Peltz and M. Weiss, "State and Local Government Initiatives for Economic Development through Technological Innovation," Journal of the American Planning Association 50, no. 3 (Summer 1984):270-79.

the capital can be used only for new and expanding businesses engaged in production, warehousing, or research and development. Some states use IRBs and GOBs for quite specific purposes: in Ohio, for example, minority-owned construction companies qualify for special IRBs. States differ, also, in the interest rates they allow local governments to pay on the bonds. Most states let the market dictate the interest rate. Because interest from most state and locally issued IRBs and GOBs is not currently taxed by the federal government, the interest rate paid can be several points below the comparable private rate. This savings is passed on to the corporate recipient of bond revenues. Some states, however, cap the interest rate on IRBs and GOBs: in California and Nevada, for example, local governments cannot pay more than 12 percent on issues. When interest rates elsewhere are high, this cap reduces investor demand for IRBs and GOBs. In 1981, for example, only $53 million in bonds were sold by local governments in Nevada, and $77.35 million were sold in California. In Pennsylvania in that same year $2.6 billion in IRBs were sold, but at higher (adjusted) market rates of 14 to 17 percent.[41]

A large number of states also reduce or excuse taxes for certain types of industries or industrial property. Thirty-two either mandate local governments to abate property taxes, or allow abatement as a local option. In several cases (Connecticut, Michigan, Mississippi) this abatement is tied to enterprise zone legislation and thus applies only to businesses in designated areas. In other cases abatement applies only to certain industries or types of facilities. In New Hampshire, for instance, only new railroads and small power facilities are excused from property taxes; in North Dakota, only new industries and speculative buildings; in Oklahoma, only renovated structures; in Colorado, only forest lands; in Arkansas, only cotton mills financed with IRBs; in Maine, only mining properties. Property tax abatement legislation applies generally, or to all new or expanding businesses, in several states (Florida, Idaho, Louisiana, Mississippi, Montana, Rhode Island, South Carolina, South Dakota). Thirty-six states exempt business inventories in full or in part from local or state property (income) taxes. The extent of that exemption varies markedly. In Idaho business inventories are fully exempt from all taxation within the state; in North Carolina, on the other hand, an income tax credit is allowed for local inventory taxes paid on raw materials and goods in the process of manufacture that are considered "excessive." This clearly is not as beneficial as the Idaho law. There is likewise considerable variance among states that exempt goods in transit (forty-one states), fuels and materials (forty-three), and industrial equipment (forty-four).

The widespread use and diversity of state and local tax abatements and exemptions makes reliable interstate comparisons of statutory income and sales tax rates especially difficult. Of course, states that do not tax corporate income (Nevada, North Dakota, Wyoming) or sales (Alaska, Montana, New Hampshire, Oregon) have zero percent effective income or sales tax rates. But other states' effective tax rates do not necessarily coincide with their statutory rates. Oregon and Vermont, for example, both had a marginal corporate income tax rate of 7.5 percent in 1981.[42] But because Oregon allowed fewer exemptions and provided no tax credits in that year, its effective tax rate was significantly higher. Similarly, taking exemptions and credits into account, North Carolina's 6 percent statutory tax seems to be less burdensome on businesses than Kentucky's 6 percent tax.

The third widely used program shown in table 1 is manpower training (thirty-eight states). As with the incentives discussed above, states differ substantially in the organization and size of such programs. Some states provide pre-employment instruction themselves, in community and technical colleges, for example, either to a general group of qualified individuals or to individuals preselected by resident businesses (as North Carolina does). Those and other states may also share with businesses the cost of on-the-job training, not by paying trainees' wages, but by providing materials and/or allowing tax credits (Connecticut, Indiana, Missouri, New York, Pennsylvania). These differences in approach affect states' expenditures on training and also the number of workers trained and placed in permanent jobs. The North Carolina program, discussed above, is among the largest in the country; the state's share exceeded $2.2 million in fiscal 1982 for the instruction of 5,819 trainees for 80 companies. Ohio ($3.5 million for fiscal 1982, approximately 7,000 trainees), California ($8.3 million/year for 1980–82, 6,835 trainees in January 1982), and Massachusetts ($5 million in fiscal 1982, 5,562 trainees) also have large programs. At the other extreme, such states as Hawaii ($151,150 in fiscal 1980–81, 118 trainees specifically for the garment industry), Kansas ($32,533 in fiscal 1981–82, 72 trainees, for manufacturing companies), and Iowa ($0 in fiscal 1981–82, no trainees) have minimal programs.

Innovative Programs

Table 1 also shows a number of innovative activities, including state-funded or -chartered equity/venture capital companies (ten states), privately sponsored development credit corporations (sixteen states), and state-specific tax credits (twenty-three states for investment, nineteen states for employment, and fourteen states for research and development).

State-funded or -chartered equity/venture capital companies provide money for businesses in the form of debt and equity. Typically, these companies' initial resources come from legislative appropriations and the sale of stock to investors. Once some businesses have been helped, operating capital is generated by the repayment of loans (at market rates of interest), by the payment of royalties (in the case of Connecticut's Product Development Corporation), and by the sharing of profits when equity capital had been provided. The equity/venture capital companies differ in scope. The Alaska Resource Corporation provides loans and loan guarantees to a wide range of businesses. The companies in Indiana, Louisiana, Maine, Michigan, New York, and North Carolina focus their attention on small businesses; those in Massachusetts, New York, and Wisconsin concentrate on businesses in distressed areas; those in Connecticut, Massachusetts, New York, and North Carolina (specifically the Innovation Research Fund) direct their efforts to businesses developing new and/or technologically advanced products.

The second innovative activity included in table 1 is the establishment within states of private development credit corporations as conduits for government loans (through the Small Business Administration, for example) and/or as sources of additional loanable funds. Most of these corporations are formed using capital from financial institutions, which then participate in further loan packaging and consequently share proceeds when loans are repaid, typically at rates above prime. Some development corporations act as lenders of last resort

(Kentucky, Massachusetts), sometimes to small businesses (Georgia, Maryland, Missouri) or businesses in distressed areas (Massachusetts), and sometimes at lower than market rates (Arkansas).

The establishment of private development credit corporations within states is one way to increase states' share of Small Business Administration loans, allocated through Section 503 of the Small Business Act. Section 503 loans come from the Federal Finance Bank through SBA to Certified Development Companies that can be set up statewide. The distribution of SBA money by state, under Section 503, is quite uneven, ranging from over $22 million in California to zero in six states in 1983 (see table 2). Of the sixteen states that had development credit companies in 1983, only one did not receive Section 503 monies. Several smaller states (Iowa and Kentucky) seem to have received a disproportionate share of such funds, perhaps because they had established statewide development corporations. For 1982 and 1983 combined, North Carolina ranked nineteenth in Section 503 funds received, even though it was the tenth largest state by population. Table 2 also shows the distribution of SBA loans to local development companies, under Section 502. The variance is large there, as well. North Carolina received no funds under this program in 1982 or 1983.

The third innovative activity included in table 1 is the use of state-specific tax benefits, either to induce firms in general to make new or additional investments in the state and to hire new workers, or to encourage particular activities. The general tax incentives include investment and employment tax credits, designed after similar programs at the federal level. The general investment tax credit rate varies, from 100 percent for select businesses in North Dakota, to 0.5 percent in Colorado, Illinois, Kansas, and Oklahoma. The general employment tax credit varies from 25 percent of wages in Florida to 1 percent in Montana and North Dakota, and from $50 per new job created in Colorado and Kentucky to $1,500 per new job (in the first year) in Maryland.[43]

Tax incentives intended to promote particular kinds of behavior take several forms. Fourteen states reduce tax liability for businesses engaged in research and development. Seven states that have enterprise zone legislation, and three other states (Massachusetts, Pennsylvania, Virginia), use tax credits and deductions as incentives for businesses to operate in designated locations. As previously noted, some states use these programs to reward firms for training workers or for hiring workers of a particular type (California, Massachusetts). Others use credits or deductions to reward taxpayers for having contributed to community assistance projects (Delaware, Florida, Wisconsin), car

Table 2. Gross Amount of Loans Approved to Development Companies ($millions)

State (1980 pop. rank)	§ 502		§ 503	
	1982	1983	1982	1983
Alabama (22)	--	--	.220	2.729
Alaska (50)	--	--	--	--
Arizona (29)	--	.225	.895	3.220
Arkansas (33)	--	--	--	--
California (1)	8.138	6.146	19.067	22.848
Colorado (28)	.836	.082	.213	.591
Connecticut (25)	--	--	.380	.875
Delaware (47)	--	--	--	.596
Florida (7)	.615	--	1.268	.773
Georgia (13)	1.520	2.351	1.152	5.331
Hawaii (39)	--	--	.202	.653
Idaho (41)	--	--	--	--
Illinois (5)	--	1.651	1.316	8.099
Indiana (12)	--	--	.168	1.192
Iowa (27)	.108	--	1.043	2.390
Kansas (32)	--	.089	.337	1.116
Kentucky (23)	.193	--	1.580	5.230
Louisiana (19)	--	--	1.899	1.283
Maine (38)	--	--	2.009	2.076
Maryland (18)	--	--	1.732	2.812
Massachusetts (11)	1.490	.952	6.780	10.688
Michigan (8)	.555	--	6.432	9.837
Minnesota (21)	--	--	.420	2.169
Mississippi (31)	--	--	--	1.765
Missouri (15)	.114	.166	3.779	7.647

Note: The gross amount equals SBA amount in every case but one for § 503, and exceeds SBA amount in most cases for § 502.

pooling (California, Colorado), or child care programs (Ohio). And some use tax incentives specifically to increase the liquidity of small businesses (Colorado, Montana, North Dakota).

Targeting

The last three columns of table 1 show, among other things, which states target programs to distressed geographic areas, small businesses, and "favored" industries. Targeting is recorded in those columns if any of the programs used by a state has an *explicit* geographic, small business, or industry-specific component. Targeting is not indicated for

Table 2, continued

State (1980 pop. rank)	§ 502		§ 503	
	1982	1983	1982	1983
Montana (44)	--	--	.043	.173
Nebraska (35)	.081	.135	.020	.781
Nevada (43)	--	--	--	.184
New Hampshire (42)	--	--	--	.198
New Jersey (9)	--	--	--	1.647
New Mexico (37)	--	--	--	--
New York (2)	1.440	.845	3.696	8.162
North Carolina (10)	--	--	.284	3.663
North Dakota (46)	.153	--	.220	.114
Ohio (6)	--	--	7.823	17.215
Oklahoma (26)	--	--	.096	1.851
Oregon (30)	--	--	.046	4.029
Pennsylvania (4)	--	--	.696	2.082
Rhode Island (40)	--	--	3.277	3.083
South Carolina (24)	--	1.293	.101	1.722
South Dakota (45)	--	--	--	.329
Tennessee (17)	--	--	.236	1.402
Texas (3)	.385	--	4.326	13.425
Utah (36)	--	--	1.105	1.756
Vermont (48)	--	--	.942	.329
Virginia (14)	--	--	--	2.073
Washington (20)	--	--	4.152	7.851
West Virginia (34)	--	--	--	--
Wisconsin (16)	4.259	6.454	3.361	10.404
Wyoming (49)	--	--	--	--

Source: Data supplied to the author by the Small Business Administration.

states that give priority to particular geographic areas or industries in the allocation of monies under general assistance programs (as North Carolina does with its speculative building and incubator facility programs, for example).

Eighteen states target distressed geographic areas. Nine of those (Connecticut, Florida, Kansas, Kentucky, Louisiana, Maryland, Minnesota, Missouri, Ohio) have established enterprise zones, which are areas within states designated for special treatment under several state programs. Usually these zones are in distressed urban neighborhoods, but in Kansas the zones can be coterminous with community boundaries, in Louisiana zones also are set up in rural areas, and in Florida zones are set up in one hundred disparate locations. In most cases the state provides special tax credits, property tax abatement, and special loans to businesses in the enterprise zone that employ workers living within the zone. Eight other states have other types of geographic targeting. In California, the Office of Small Business Development only provides direct state loans and loan guarantees to businesses in distressed cities. Similarly, the Illinois Industrial Development Authority, the Massachusetts Community Development Finance Corporation, the New York State Urban Development Corporation, the Pennsylvania Industrial Development Authority, and the Wisconsin Community Development Finance Corporation all have geographically targeted loan programs, many at below-market interest rates. Virginia offers a 50 percent tax credit to firms that invest in distressed areas.

Twenty-one states target small businesses, using a wide variety of programs. One of the techniques for helping small businesses, not discussed above, is the use of umbrella bonds (ten states). Unlike IRBs or GOBs, which are typically issued to finance single large projects, umbrella bonds are issued to finance several smaller projects. In addition to the states that use umbrella bonds, California, Louisiana, Massachusetts, Ohio, and Pennsylvania all have aggressive small business programs.

The last type of targeting indicated in table 1 is that directed at favored industries (forty-three states).[44] Most of this targeting is done through the provision of property, sales, and income tax exemptions. For example, Alabama exempts cotton, tobacco leaf, and textile mill products and materials from the property tax. Hawaii exempts paper and pulp manufacturers from property and sales taxes, and canners of pineapple and sugar products from the sales tax. Kentucky exempts most alcohol producers from property and sales taxes. New Hampshire exempts ski areas from the property tax. Oklahoma gives favorable tax treatment to petroleum products and equipment. Texas exempts equipment used in mineral exploration and production from sales taxes. West Virginia and Washington exempt lumber and wood products from sales taxes. And North Carolina exempts tobacco, peanuts, cotton, and other agricultural products purchased by manufacturers from the sales tax. Some states also target loans and bond issues to favored industries. Alaska uses umbrella bonds for fishing, tourism, and mining alone, and directs state loans to fishing and agriculture businesses. Kentucky directs bonds to manufacturing and tourism businesses. Oregon uses loans and bonds for ports, in particular. And so on. The pattern is clear: states tend to protect their dominant industries.

The zeroes in the "favored industries" column of table 1 denote targeting of high-technology businesses in particular. The U.S. Office of Technology Assessment refers to such efforts as "dedicated high-technology development programs," which are "chartered and at least partially funded by the state government, and specifically targeted on the creation, attraction, or retention of high-technology firms."[45] This column together with the columns for "special development task force" and "research park" provides a good summary of high-technology development activity in the United States. Essentially two types of states are active in high-technology programs: (1) traditionally active progressive states, in the Midwest, Mid-Atlantic, and New England regions, that have favored strong government for a long time (Connecticut, Illinois, Indiana, Massachusetts, Michigan, New York, Ohio, Pennsylvania, Rhode Island) and (2) rapidly growing states in the Sunbelt, many of which already have a start in the high-technology area (California, Colorado, Florida, Georgia, Hawaii, Mississippi, New Mexico, North Carolina, Texas).

Critique

Using information from the preceding discussion we can characterize North Carolina's overall development strategy as follows:

- North Carolina relies more heavily on programs to recruit outside industries, especially high-technology businesses, than on programs to help make existing traditional businesses technologically more advanced, or to promote the growth of emerging resident businesses.
- Compared to other states, North Carolina has large manpower training programs, both for workers destined for particular jobs (via Customized Industrial Training) and for the general population (via community colleges and technical schools). It places considerably less emphasis on job *re*training and referral, even though there is a high incidence of plant closings in the state.
- North Carolina provides relatively large plant subsidies through its speculative building and incubator facility programs.
- North Carolina keeps its effective tax rate lower than most other states by allowing numerous deductions and exemptions from a relatively low statutory corporate tax. Local property taxes are

also low in the state. Because its state–local tax burden is so low (forty-seventh among states), North Carolina has not needed to use as many targeted tax credit programs as other states.

- North Carolina provides relatively little capital to businesses directly. It does not issue IRBs or GOBs itself and its local governments float a relatively small volume of those bonds. There is no mechanism for guaranteeing private loans at the state or local level, so the volume of bank financing is less than it could be. Similarly, there are no private development credit companies in the state to generate loanable funds, and the new state equity corporation has very little capital. In addition, the state has resisted using its pension fund for development purposes.[46]
- North Carolina lags behind other states in efforts to help distressed areas.

Some specific analytic questions regarding this strategy can be posed. (1) Are specific incentives needed to recruit firms in general, or of particular types, if taxes and wages are already low? (2) Are direct financial assistance programs necessary if economic conditions and state policy ensure firms of adequate capital for investment? And (3) is it necessary for states to help develop small, indigenous businesses and distressed regions if the overall state economy is strong? On the first of these questions, accurate comparisons between North Carolina's development efforts and those in other states are difficult to make. North Carolina's relatively limited use of tax credits, for example, should not be interpreted as a failure of policy, but as a reasonable response to existing conditions. This applies, as well, to the second question. North Carolina may not need to be as aggressive as other states in the provision of financial capital, because low costs and vibrant markets make the state an appealing site for many businesses regardless of the limited availability of state-financed or state-backed loans. That does not apply to all businesses, however. North Carolina has a large number of firms that do not have sufficient liquidity to grow and that cannot secure private bank loans. The limited scope of existing capital programs hurts those businesses in particular. This observation helps to answer the third question. Targeted programs are needed, even in strong economies, to help businesses and regions that have particular problems. But there is considerable debate in the literature on industrial policy about whether such targeting creates a drain on the overall economy or is ultimately stimulative.[47]

Further consideration of these large, analytic questions is beyond the scope of discussion here. The critique that follows is much more limited; it is simply an assessment of whether North Carolina's recruitment-oriented development strategy is likely to achieve, in a satisfactory way, the five policy goals enumerated early in this chapter. An overriding issue is whether industrial recruitment induces businesses to relocate. Hekman and Greenstein's research (chapter 6, this volume) suggests that most of the companies that moved facilities to North Carolina during the past several years would have done so even if they had not been contacted by state officials. That is not surprising, because most such plants opened in the state are branches of large firms that typically employ experts to collect and analyze information on alternative sites. If state recruiters are unnecessary, the $4 million (or more) spent biennially to employ them could be put to better use.

Consider the *balanced development* goal first. The only direct mandates for geographic targeting in state policy are (1) a stipulation in the new small business legislation (incubator facilities and the innovation research fund) to use local unemployment rates as one criterion for disbursing money, and (2) a provision in the speculative building program to locate some of those facilities in distressed communities. The first of these is not likely to have a significant effect on development patterns, because the small business programs are so small in size. The second type of targeting may seem to have worked; almost half of the speculative buildings erected by 1980 were in depressed counties. But it is fair to assume that the immigrating businesses would have located in those places even without a plant subsidy program. Most economists agree that proximity to markets, appropriate labor, and raw materials outweigh other considerations in firms' location calculus.[48] A broader implication of this economic rationale is that most high-technology facilities coming into North Carolina or starting anew will continue to concentrate in the Piedmont Crescent stretching from Raleigh to Charlotte, especially in the Raleigh–Greensboro corridor, near existing high-technology companies, major universities and airports, and the Microelectronics and Biotechnology Centers in the Research Triangle Park. In fact, by placing the research centers in the park and promoting the park as a choice location, the state actually has been undermining its goal of balanced growth.

North Carolina's recruitment-oriented strategy may also violate the *job stability* goal, because the facilities being sought are branches of large multiplant, multilocational firms with headquarters in other states and countries. These firms will typically search beyond the borders of North Carolina for future profit opportunities. A high-technology company based in California, for example, may decide to use the profits from a North Carolina plant to finance the construction of new high-technology plants in Mississippi or Thailand, or to buy another kind of company altogether. In time, the firm may close the North Carolina plant, with a resultant loss of jobs. This long-term outcome of capital mobility is more than hypothetical.[49] State officials agree that North Carolina would do better to cultivate headquarters operations than branch plants, but they have not addressed the problem that companies are not as inclined to relocate central offices.[50] A stronger emphasis on small, locally owned businesses would be better for the state in this regard.

Though North Carolina's overall strategy is likely to violate its goals of balanced growth and stability, it may well achieve the goals of *job creation* and *wage growth* (assuming that recruitment works), but not in a satisfactory way. Previous study of the microelectronics industry in North Carolina has detected two latent pitfalls.[51] (1) The demand by new facilities for skilled workers will far outstrip the supply of such workers for the foreseeable future, even accounting for the new education programs discussed above. This will be true especially within the seven-county "projected location zone" shown in figure 1. New plants will thus be forced to import skilled workers from other states—as existing firms have already done[52]—or induce those who are already employed in the zone to switch jobs. Competitive bidding for engineers, managers, and skilled machinists already employed in the area may raise those workers' pay, thus increasing the state's average manufacturing wage rate; but those wage benefits would not necessarily filter down to less skilled workers or to workers in outlying counties where wages are particularly low. Higher average wage rates could actually be accompanied by greater inequality of income. (2) There also will be excess demand for semiskilled workers within the projected location zone, because there is a dearth of workers that already have some technical skills, and of workers who can be trained. The initial burden of this shortage will probably fall on businesses already located in the zone; semiskilled workers can be pirated away, and the prospect of higher wages can induce unskilled workers to undertake job training. To retain both groups of workers, existing companies would have to increase pay scales. That could drive marginal firms out of business.

These projected outcomes for the labor market invite an interesting interpretation. Planners in North Carolina and other southern states are faced with the problem of raising average wage rates without damping business investment. If they are successful in raising wage rates uniformly, the incentive for new firms to locate in those states will be diminished and the health of the existing firms in traditional industries that rely on low-wage workers may be jeopardized. The strategy policy-makers in North Carolina have adopted gets around this dilemma, unwittingly or otherwise. As high-technology firms move into the state, particularly into a circumscribed geographic area, the average manufacturing wage rate for the state will rise. But because those wage increases will not be spread uniformly to workers of all skills and in all locations, low-wage workers will remain available for employment in traditional low-skilled industries. From the industrial recruiters' perspective this mechanism for increasing average wage rates is far superior to the solution of unionization, which is well known to discourage business immigration. From the perspective of low-wage workers in outlying counties, however, the state would do better to assume a neutral position on unionization, perhaps by repealing its right-to-work laws; unionized workers in some industries receive up to 35 percent more pay than their nonunion counterparts doing similar work.[53]

Finally, there is convincing economic evidence that the mix of specific programs used in North Carolina will not prove *cost effective.* A number of recent studies sponsored by the Urban Institute have ranked specific economic development programs according to their cost-to-government/benefit-to-firm ratios.[54] Costs include direct state outlays and forgone tax revenues over the programs' life, under realistic economic assumptions. Benefits are calculated as the increase in firms' asset values due to incentives. That is a valid measure of benefits for financial assistance programs whose purpose is to induce more investment and employment by increasing firms' liquidity. The Urban Institute's effectiveness ratio is thus a measure of the public cost per dollar of increased private liquidity. Table 3 summarizes the results of these studies. For all types of firms analyzed, industrial revenue bonds (IRBs) and loan guarantees are cost-effective. For both small/new and existing/expanding firms (presumably, at any level of profitability) direct loans and interest subsidies are also cost-effective. North Carolina does not rely heavily on these types of financial assistance programs. Equity programs are rated cost-neutral. North Carolina's new Innovation Research Fund is an equity program, but as noted above, it is poorly funded. Plant subsidies—for speculative buildings and incubator facilities, for example—and worker training programs are not cost-effective, though the results for worker training may not apply to high-technology businesses.[55] North Carolina relies heavily on both. The table does show that some cost-*in*effective programs are *not* used in North Carolina, including equipment subsidies and site preparation.

Table 3. Summary of Programs' Cost-effectiveness

Type of Business	Cost-Benefit Category		
	Cost-effective	Neutral	Not Cost-effective
Low profit	IRBs	Equity sharing	Plant subsidies Equipment subsidies Worker training[a] General tax abatemen
High profit	IRBs	Equity sharing	Plant subsidies Equipment subsidies Worker training[a] General tax abatemen
New, small[b]	IRBs Direct interest subsidies Loan guarantees Direct loans	Equity sharing	Site preparation Plant subsidies
Existing, expanding[b]	IRBs Direct interest subsidies Loan guarantees Direct loans	Equity sharing	Site preparation Plant subsidies

[a]For high-technology and traditional businesses

[b]Based on study of Louisiana

Sources: William W. Hamilton and L. Ledebur, *Options for Industrial Finance* (Washington, D.C.: Urban Institute, 1983); David W. Rasmussen, M. Bendick, Jr., and L. Ledebur, *The Cost Effectiveness of Economic Development Incentives* (Washington, D.C.: Urban Institute, 1982); Urban Institute and Coopers & Lybrand, *Recommendations for a Cost Effective Business Incentive Program for Louisiana* (Washington, D.C.: Urban Institute, October 1983).

Conclusion

North Carolina has shown considerable initiative in its high-technology development policy. But by themselves, the few visible programs for which it has received recognition are not likely to fulfill the promise of economic prosperity for all state citizens. The strategy relies too much on industrial recruitment and too little on programs for existing traditional and emerging resident businesses. Moreover, the mix of programs used is not cost-effective.

These findings lead to a number of policy recommendations. As a general matter, *North Carolina needs to broaden its menu of develop-*

ment programs. Because the state has varied goals for development, it needs a varied set of programs. The importance of comprehensiveness has been stressed in several recent studies. The National Governors' Association, for instance, concluded that "experience to date indicates that multifaceted, comprehensive approaches provide the best opportunity for actualizing state potential for technological innovation."[56] An Urban Institute report to the Louisiana Department of Commerce firmly recommended that "the state have available a flexible array of incentives to respond to the needs of the different firms it seeks to assist."[57] To make its strategy more comprehensive, North Carolina must include more programs for existing traditional businesses, existing small enterprises, and emerging new companies. But those programs must be selected carefully to be consistent with the state's development goals. Thus *North Carolina needs to reexamine the programs it now uses, and evaluate new programs it is considering, to ensure that they can meet the cost-effectiveness, employment, and balanced growth goals that have been articulated.* State planners seldom undertake this type of careful analysis.[58]

Information presented above suggests specific programs that may be useful in North Carolina. Some of these have already been proposed by Governor Hunt's Task Force on Science and Technology. In particular, *policy-makers in North Carolina should consider having the state issue IRBs and umbrella bonds itself, guaranteeing loans, and providing direct loans and interest subsidies to the businesses it seeks to assist*. Such programs are cost-effective and permit targeting of declining traditional businesses and small enterprises, especially those investing in depressed areas. Finally, *policy-makers should reevaluate the state's stance on unions, and perhaps repeal the right-to-work legislation*, because there is a high correlation between union representation and higher wage levels. That policy could reduce the rate of new job formation by immigrating branch plants; but if policy focused simultaneously on existing and emerging resident businesses, enough jobs could still be created to meet the state's needs.

Notes

1. For example, *N.C. Insight* 4, no. 3 (Fall 1981), a special issue on the microelectronics industry in North Carolina; *New York Times Magazine*, 20 March 1983, 32, 34, 36, 66, 69; National Governors' Association, *Technology and Growth: State Initiatives in Technological Innovation* (Washington, D.C., October 1983).

2. These documents include James B. Hunt, Jr., "An Economic Development Strategy for the 1980s" (Mimeo, Raleigh: Office of the Governor, September 1980); idem, State of the State Address (Mimeo, Raleigh: Office of the Governor, January 1983); Governor's Task Force on Science and Technology, report, vol. 1: *New Challenges for a New Era: A Vision of Opportunity*, and vol. 2: *Economic Revitalization through Technological Innovation* (Raleigh: North Carolina Board of Science and Technology, November 1983); and North Carolina Balanced Growth Policy, Chapter 412 of the 1979 Sessions Law (HB 874), Session 7(1), 19 April.

3. Based on a conversation with Jack Hawke, Special Assistant to Governor Martin for Policy, 22 February 1985.

4. U.S. Department of Commerce, Bureau of the Census, *Annual Survey of Manufactures, 1979* (Washington, D.C., 1982).

5. Hunt, State of the State Address.

6. See Edward Bergman, "Local Economic Development Planning in an Age of Capital Mobility," *Carolina Planning* 7, no. 2 (1980): 29–37.

7. Governor's Task Force on Science and Technology, *Economic Revitalization*, 68.

8. North Carolina Balanced Growth Policy (1979).

9. Edward Gramlich and E. Laren, "The New Federalism," in *Setting National Priorities: The 1983 Budget*, ed. Joseph Pechman (Washington, D.C.: Brookings Institution, 1982), 151–87.

10. The full titles of these are the 1981 Economic Recovery Tax Act and the 1982 Tax Equity and Fiscal Responsibility Act. The estimate is from Citizens for Tax Justice, *Tax Policy Guide No. 1: How the States Can Respond to the 1981 Changes in Federal Depreciation Rules* (Washington, D.C., 26 November 1982).

11. North Carolina Center for Public Policy Research, *Federal Budget Cuts in North Carolina, Part II* (Raleigh, April 1982).

12. Based on data from an interview with Harvey Lincoln of the North Carolina Office of Budget and data from Citizens for Tax Justice, *Tax Policy Guide No. 1*.

13. Governor's Task Force on Science and Technology, *New Challenges*, 18.

14. James C. Cobb, *The Selling of the South: The Southern Crusade for Industrial Development, 1936–1980* (Baton Rouge: Louisiana State University Press, 1982).

15. The *Raleigh News and Observer*, 20 October 1979, 8, as cited in Thomas Vass, "Industrial Recruitment and the Path of North Carolina's Economic Development in the Year 2000" (Mimeo, Raleigh: N.C. Department of Labor, 1982), n. 1.

16. Vass, "Industrial Recruitment," reports that in 1979 and 1980, 380 private development organizations initiated 197 of the 306 new projects begun in North Carolina.

17. John M. Kline, *State Government Influence in U.S. International Economic Policy* (Lexington, Mass.: Lexington Books, 1983), 57–58.

18. See Government Research Corporation, *High Technology: Public Policies for the 1980s*, A *National Journal* Issues Book (Washington, D.C., 1983), 111 (quoted); Alexander Grant & Co., *A Study of Manufacturing Business Climates in the 48 Contiguous States of the U.S.* (Washington, D.C., 1980); "More Elbow-room for the Electronics Industry," *Business Week*, 10 March 1980, 94–106; and Dale Whittington, chapter 1, this volume.

19. Based on data provided by the North Carolina Department of Commerce, Industry Financing Division, "Urban Development Action Grants" (Mimeo, Raleigh, March 1983).

20. From North Carolina Department of Commerce, Industry Financing Division, "Basic Building Program" (Mimeo, Raleigh, 1979).

21. N.C. Department of Commerce, "Urban Development Action Grants."

22. Urban Institute et al., *Directory of Incentives for Business Investment and Development in the United States* (Washington, D.C.: Urban Institute, 1983).

23. Urban Institute et al., *Directory of Incentives*, 468.

24. Hunt, State of the State Address.

25. Governor's Task Force on Science and Technology, *Economic Revitalization*, 7.

26. The specific function of the Microelectronics Center is described by George Herbert, "Why We Need the Microelectronics Center," *N.C. Insight* 4, no. 3 (Fall 1981); see also Governor's Task Force on Science and Technology, *Economic Revitalization*, 17 (quoted).

27. For complete details see Governor's Task Force on Science and Technology, report, vol. 3: *Research and Higher Education: Building a Foundation for Innovation* (Raleigh: North Carolina Board of Science and Technology, November 1983).

28. Governor's Task Force on Science and Technology, *Economic Revitalization*, 57.

29. Catherine Armington, C. Harris, and M. Odle, "Formation and Growth in High Technology Business: A Regional Assessment" (Washington, D.C.: Brookings Institution, 30 September 1983), unpublished typescript.

30. Governor's Task Force on Science and Technology, *Economic Revitalization*, 67.

31. Hunt, State of the State Address.

32. The modernization of Burlington Industries' Erwin plant is a frequently cited example. See Governor's Task Force on Science and Technology, *Economic Revitalization*, 67–70, and National Governors' Association, *Technology and Growth*, 1.

33. Governor's Task Force on Science and Technology, *Economic Revitalization*, 70–71, 73–77.

34. Based on calculations for the South and the United States in Armington, Harris, and Odle, "Formation and Growth."

35. My estimate is based on mean employment estimates for plants in different size categories.

36. See John Rees, "Regional Industrial Shifts and the Internal Generation of Manufacturing in Growth Centers of the Southwest," in *Interregional Movements and Regional Growth*, ed. W. C. Wheaton (Washington, D.C.: The Urban Institute, 1981); John Rees and H. Stafford, *A Review of Regional Growth and Industrial Location Theory: Towards Understanding the Development of High Tech Complexes in the U.S.* (Washington, D.C.: U.S. Office of Technology Assessment, May 1983).

37. See David Birch, *The Job Generation Process* (Cambridge, Mass.: MIT Program on Neighborhood and Regional Change, 1979); idem, "Who Creates Jobs?" *The Public Interest* 35 (Fall 1981); Roger Schmenner, *The Location Decision of Large Multi-plant Companies* (Cambridge, Mass.: Harvard–MIT Joint Center for Urban Studies; Barry Bluestone and B. Harrison, *The Deindustrialization of America* (New York: Basic Books, 1980); E. A. Erickson, "Corporate Organization and Manufacturing Branch Plant Closures in Non-metropolitan Areas," *Regional Studies* 14, no. 6 (1980): 491–503; David L. Barkley, "Plant Ownership Characteristics and the Locational Stability of Rural Iowa Manufacturers," *Land Economics* 54, no. 1 (1978): 92–99; P. N. O'Farrell, "An Analysis of Industrial Closures: The Irish Experience 1930–1973," *Regional Studies* 10, no. 4 (1976): 433–49; and R. A. Henderson, "An Analysis of Closures amongst Scottish Manufacturing Plants between 1966 and 1975," *Scottish Journal of Political Economy* 27, no. 2 (1980): 2–74.

38. Governor's Task Force on Science and Technology, *Economic Revitalization*, 91–93.

39. The table must be interpreted with care, for several reasons. First, no adjustment has been made for program size or quality. For example, New Hampshire has a local issue IRB program but no bonds were actually sold in 1981; and as indicated in earlier discussion above, North Carolina's state-funded equity capital and small business programs are small and untested. Nonetheless, those programs are noted in the table. Second, some programs do not feature in the appropriate place in the table. For example, twenty states are shown to have direct loan programs, but eleven others make similar loans through private development credit or equity/venture capital companies. Conversely, double counting occurs because some of the categories listed along the top of the table overlap. For example, business procurement assistance may be targeted specifically for small businesses, in which case an X appears in both columns. A related problem is that some categories in the table are mutually exclusive. For example, states without corporate income taxes cannot have tax credits and deductions, and those without sales/use taxes cannot have tax exemptions. Third, the table is not comprehensive. Some state programs have been omitted altogether. In addition, city- and county-sponsored programs are not included, even though many have the same purposes as the state programs. In its survey of state legislative climates, *Industrial Development* (see fifteenth and sixteenth Annual Reports in the March–April 1981 and January–February 1982 editions) reports, for example, that twenty-six states have locally sponsored speculative building programs; in fifteen states local governments pro-

vide free land to businesses; sixteen states have city or county business loan programs; nineteen have city or county loan guarantees; and so on. Moreover, errors during collection of the original data may have led to oversights concerning programs that are reported here. Because of these limitations we cannot conclude, for example, that a state with fifteen programs noted is more active in economic development than a state with five—whose programs may be larger in size and impact. Or double counting or omissions (for instance, of city- or county-sponsored programs) may distort the total program counts.

For a fuller discussion of the difference between effort and outcomes see Michael I. Luger, "Does North Carolina's High Tech Development Program Work?" *Journal of the American Planning Association* 50, no. 3 (Summer 1984): 280–90; and idem, "Explaining Differences in the Use and Effectiveness of States' Industrial Policies" (Working paper, Institute of Policy Sciences and Public Affairs, Duke University, Durham, N.C., 1985).

40. Industrial revenue bonds are repaid using revenues generated by the recipient of loan funds. General obligation bonds are backed by the full faith and credit of the issuing government. Umbrella bonds are like general obligation bonds but are issued to cover a number of small projects.

41. The interest rates were quoted by Jeff Burt of the Montgomery County, Maryland, Department of Housing and Community Development, interviewed 27 September 1983. The tax-free treatment of most state and local bonds could be curtailed under federal tax reform proposals.

42. Council of State Planning Agencies, *Book of the States* (Lexington, Ky.: Hall of the States, 1983), 402.

43. Some other states calculate investment and jobs credits as a percentage of those credits provided by the federal government.

44. The list omits industries that are widely favored in state tax policy—for example, industries that use trucks and water craft.

45. *Technological Innovation and Regional Economic Development* (Washington, D.C., 1983), 7.

46. The state had assets worth $6 billion in its pension fund in 1983. Managers of the fund and some state employees oppose the use of these assets for development, fearing that any restrictions placed on investments or loans would reduce the overall yield. That has not occurred in states that have tapped such funds, however. There is a growing literature on this; for example, see D. A. Smart, "Investment Targeting: A New Direction for Pension Funds," *Commentary* 4, no. 3 (July 1980), 7–10.

47. This literature is now extensive. See, for example, Ira Magaziner and R. Reich, *Minding America's Business* (New York: Random House, 1982), and *Challenge* 26, no. 3 (July–August 1983).

48. See, for example, Bennett Harrison and S. Kanter, "The Political Economy of States' Job Creation Business Incentives," *Journal of the American Institute of Planners* 44, no. 4 (1976): 424–35; and Roger Vaughan, "State Tax Incentives: How Effective Are They?" *Commentary* 4, no. 1 (January 1980), 3–5.

49. See Bluestone and Harrison, *The Deindustrialization of America.*

50. At least according to a recent survey of microelectronics executives; see Annalee Saxenian, "Silicon Chips and Spatial Structure: The Industrial Basis of Urbanization in Santa Clara County, California" (Working paper 345, Institute of Urban and Regional Development, University of California, Berkeley, 1981).

51. Michael I. Luger, "Promises and Policies: The Economic Hope of the Microelectronics Industry," *N.C. Insight* 4, no. 3 (Fall 1981): 27–32; and idem, "Does North Carolina's High Tech Development Program Work?"

52. The General Electric Corporation, for example, had to relocate more than 100 skilled workers to staff positions in their Research Triangle Park facility.

53. See Michael I. Luger, "A Note on the Relationship among Unionization Rates, Wages and Business Migration" (Working paper, Bureau of Business and Economic Research, University of Maryland, College Park, January 1984); and U.S. Department of Labor, Bureau of Labor Statistics, *Industry Wage Survey: Semiconductors, September 1977,* Bulletin 2021 (Washington, D.C., 1979).

54. These include William W. Hamilton and L. Ledebur, *Options for Industrial Finance* (Washington, D.C., 1983); Phyllis Levinson et al., *Industrial Incentives in Local Economic Development* (Washington, D.C., September 1981); David W. Rasmussen, M. Bendick, Jr., and L. Ledebur, *The Cost Effectiveness of Economic Development Incentives* (Washington, D.C., January 1982); and Urban Institute and Coopers & Lybrand, *Recommendations for a Cost Effective Business Incentive Program for Louisiana* (Washington, D.C., October 1983).

55. See Armington, Harris, and Odle, "Formation and Growth."

56. National Governors' Association, *Technology and Growth,* 105.

57. Urban Institute and Coopers & Lybrand, *Recommendations,* 13.

58. According to Rasmussen, Bendick, and Ledebur, *Cost Effectiveness,* 1, "Given the extensive use of . . . incentives, the existing level of knowledge about the comparative effectiveness of alternative forms of industrial incentives in achieving public objectives and their costs to government is grossly inadequate." A similar conclusion was reached by the Office of Management and Budget, *Managing Federal Assistance in the 1980s* (Washington, D.C., 1980).

Microelectronics and Economic Development in North Carolina

Harvey Goldstein and
Emil E. Malizia

In discussions of national industrial policy, high-technology industry is treated as a strategically important means of increasing expected profitability, domestic job generation, or international market share. The development of high-technology industries has been proposed as the linchpin of a strategy to revive the U.S. economy.[1] What began in the late 1970s as reindustrialization—the concern for rebuilding "basic" industries (such as automobile, steel)—has become a question of stimulating and assisting "growth" industries and encouraging the industrial application of electronic, biological, and other new technologies. Although the precise definition of high-technology industries remains vague, they are popularly viewed as providers of good, stable jobs without heavy pollution, expensive infrastructure requirements, intensive energy use, or militant unions.[2]

At the state and local levels, cuts in federal expenditures, high borrowing costs, and flagging area economies have prompted public officials to seek more sustaining industrial development strategies beyond traditional "smokestack chasing." Most states and many cities are beginning to target industrial development efforts on high-technology industries in the hope of expanding local jobs and tax bases.[3] Unfortunately, as more and more jurisdictions climb onto the high-technology bandwagon, their individual efforts are bound to become less successful and more expensive.[4]

North Carolina was one of the first states to articulate a high-technology industrial development strategy focused on recruitment within the microelectronics industry. The Microelectronics Center of North Carolina (MCNC) has been established to provide a locus for state efforts. While pros and cons of this policy have been much discussed, very little attention has been directed to the unique implications of microelectronics growth for the development of North Carolina's economy. Will new patterns of trade emerge among the state's industries? Will there be unique impacts on local labor markets? Will the financial condition of affected local governments improve or deteriorate? Will new business spin-offs result from microelectronics growth?

North Carolina's industrial structure is becoming more diversified. Although manufacturing output and employment remain concentrated in traditional sectors such as textiles, apparel, and furniture, there has been significant growth in metals, machinery, and electronics manufacturing over the past fifteen years.[5] To compare growth in microelectronics appropriately to the industrial development trends unfolding in the state, several particular features of the microelectronics industry must be taken into account. Obviously the industry is "different" because of the "high-technology" content of its products and processes and because of its impressive growth rates; these are the features that make microelectronics the target of state recruitment efforts in the first place. Yet the industry may also generate unique effects due to other characteristics, such as its organizational structure, patterns of trade, labor requirements, financing requirements, and structure of ownership and control. The task in the following sections of this paper is to suggest the connections between these industry characteristics and the state's economic characteristics in order to distinguish the probable economic effects of growth in microelectronics from industrial patterns and trends that are already current.

There are established macroeconomic methods for estimating the regional economic effects of exogenous economic trends or specific events.[6] In recent years, econometric models (including input-output models) have been used to trace the effects of exogenous economic change on local levels of income, employment, output, and government revenues and expenditures. In the welfare economics tradition, cost-benefit studies have been used to assess the impacts of local projects, on both an ex ante and ex post basis.[7] Each method has its advantages and disadvantages. Econometric models take a broader perspective in predicting direct and indirect (multiplier) effects, but only for economic activities measured as output, income, or employment. Cost-benefit analysis takes a narrower, project-specific view but examines all potential economic effects (price, quantity, income, wealth).[8] Both traditions contribute to the methods of analysis used in this study.

Our analysis is comparative and qualitative in character. Two assumptions anchor our comparative strategy: (1) that in the absence of any special effort to attract the microelectronics industry to North Carolina, the state's economy will continue to grow and its industrial structure will continue to change following existing trends (see Hekman and Greenstein, chapter 6, this volume); and (2) that North Carolina's efforts to attract microelectronics firms will be successful over the next several years. On this basis qualitative analysis can distinguish potential effects of growth in the microelectronics industry from projected effects of existing statewide development trends.

Our qualitative strategy departs from traditional economic impact analysis in several ways. Rather than estimate empirically the aggregate economic effects of a projected increment of microelectronics growth, we chose to examine the special features of the industry and the state's economy reciprocally. Instead of producing quantitative simulations of future trends, we developed qualitative scenarios to capture the essential features of alternative possibilities. These scenarios are necessarily exploratory and speculative. Because of our concern with qualitative effects, direct ("first-round") impacts have been given more emphasis than indirect or induced multiplier effects. In some cases we have been forced to use national data on the microelectronics industry because comparable state data do not exist. In other cases we have preferred to use national data because they are more representative of the *future* microelectronics industry in North Carolina than are current statewide data.

In the following discussion a number of topics are reviewed under four main headings: (1) the likely trade impacts of forward and backward linkages between microelectronics and other industries in North Carolina; (2) the likely impacts on local labor markets; (3) the likely impacts on patterns of public expenditures and revenue generation; and (4) the potential for formation of new companies through spin-offs from microelectronics enterprises.

Interindustry Linkages

The enhancement of interindustry linkages and trade as a strategy for national or regional economic development has been widely accepted from the varying perspectives of economic base theory, Keynesian theory, and economic development theory.[9] Though this strategy is neither necessary nor sufficient for economic development to occur, most sectoral or industrial development policies are at least implicitly based on a rationale of realizing interindustry multiplier effects.

Patterns of interindustry linkages at the regional, national, or global levels are usually described in terms of the extensiveness and structure of their interdependence, as revealed in input-output tables. For example, the degree of hierarchy in an economy may be examined by triangulating input-output tables: structural analyses of U.S., Japanese, and

Soviet input-output tables show block-triangular patterns and the existence of "universal intermediaries." Block-triangular patterns indicate relatively autonomous, highly interdependent clusters of industries. Universal intermediaries are the industries, such as transportation, communications, utilities, trade, and selected professional and business services, from which inputs are purchased by almost all other industries..

Over the past decade an extensive amount of research has been done on industrial clustering and complexes.[10] *Industrial clusters* have been defined as groups of industries bound together by technological relations of production. *Industrial complexes*, on the other hand, are segments of industrial clusters that tend to locate in the same regions or metropolitan areas. Studies using a variety of techniques of data analysis have confirmed the existence of numerous clusters and complexes, including a complex of communications and electronics industries. We have examined the interindustry linkages of the semiconductor industry, Standard Industrial Classification (SIC) 3674, first from a purely technological perspective (industrial cluster) and then from a spatial perspective (industrial complex). The findings provide key insights into the types of broader economic development that might be expected to occur in North Carolina as a result of the location of the microelectronics industry in the state.

Microelectronics and Industrial Clusters

To study the unique technological linkages of the U.S. semiconductor industry, we compared the input and output structures of that industry to those of six other industries. Three are drawn from sectors that have been important in North Carolina for some time: textiles (specifically broadwoven fabric mills and fabric finishing plants), apparel made from purchased materials, and wooden household furniture. The other three are high-technology sectors that have recently been growing in importance in North Carolina: plastic materials and resins, industrial controls, and engineering and scientific instruments. In 1982 there were several hundred establishments in each of the three traditional North Carolina industries. For the emerging industries, the number of establishments in North Carolina was much smaller: plastics, 18; industrial controls, 27; instruments, 13. There were 21 semiconductor establishments in the state in 1982.[11]

Despite the fragmentary nature of available information on interindustry linkages,[12] our findings are mutually reinforcing and consistent in revealing differences between microelectronics and comparison industries. As table 1 shows, more than 92 percent of semiconductor devices are purchased by U.S. industries on current account. The remaining output is used by the federal defense sector or represents net commodity exports. Semiconductor devices are diffused in the U.S. economy more widely than any of the comparison commodities. Only 22 percent of output is purchased by the semiconductor industry and by the top user of semiconductor devices, electronic computing equipment manufacturers. More recent information would suggest that the degree of diffusion has become even more widespread. The semiconductor industry creates a high proportion of value added compared to traditional and high-technology industries. Its inputs are more varied than three of the comparison industries and less varied than the other three.

Like the comparison industries, semiconductor manufacturers purchase goods and services from universal intermediaries. Among the most important are wholesale trade, business and professional services, eating and drinking places, real estate, and electric utility services. By far the most important manufactured goods used as inputs are other electronic components (excluding semiconductors and electron tubes), industrial chemicals, nonferrous metals, and metal-stamping products.

When the product demand of a region's industry grows, the extent of backward linkages within the region determines the magnitude of the trade multiplier effect. A regional input-output model is the usual technique employed to determine the magnitude of industry-specific trade multipliers within a state or substate economy. The high cost of constructing input-output tables has precluded the estimation of industry-specific multipliers for most states and substate regions. Recently, however, the U.S. Bureau of Economic Analysis has developed a reliable and accurate technique, RIMS II, for estimating synthetic regional input-output coefficients from national technical coefficients that are adjusted by data on the composition of regional industry (location quotients).[13] A set of industry-specific multipliers for the six comparison industries in North Carolina was generated from the RIMS II model to compare the magnitude of the output multiplier for the microelectronics industry, as shown in table 2. The multipliers reflect North Carolina's industrial composition in 1980–81. Examination of these multipliers shows that the microelectronics industry has about the same economic multiplier effect in the state economy as the other high-technology industries. All three of the traditional industries of North Carolina, however, have substantially larger economic multiplier effects. One principal explanation for this result is that the backward interindustry linkages that lead to larger trade multipliers have

Table 1. Summary of Interindustry Linkages of Semiconductors and Other Sectors, 1972 (U.S.)

	Semiconductors	Textiles
Total final demand	7.6	4.1
Total intermediate use (TIU)	92.4	95.9
Total commodity use	100.0	100.0
Percentage TIU purchased by:		
• the industry plus top user	21.5	74.1
• top five manufacturing users	56.9	85.9
• top ten industrial users	84.2	90.8
Total value added	56.4	29.3
Total intermediate inputs (TII)	43.6	70.7
Total industry inputs	100.0	100.0
Percentage TII purchased as:		
• the commodity plus top commodity input	30.3	55.2
• top five manufactured commodities	50.5	72.1
• top ten commodity inputs	69.0	83.6

Source: U.S. Department of Commerce, The Detailed Input-Output Structure of the U.S. Economy, 1972 (Washington, D.C., 1979).

had time to develop for the traditional industries but not for the high-technology industries, which have only recently become a presence in North Carolina. Over time, as the high-technology industries grow in the state, one might expect the magnitude of the multiplier to increase as more extensive backward linkages develop. Yet the nature of the type of corporate organization and strategies associated with firms in high-technology industries that are attracted to North Carolina, as we shall discuss below, may inhibit the development of backward linkages as extensive as those in the textiles, apparel, and furniture industries in North Carolina.

In summary, analysis of technological linkages leads to the conclusion that the microelectronics industry indeed has some unique interindustry linkages compared to both traditional industries concentrated in North Carolina and a sample of other high-technology industries. The magnitude of its economic multiplier effect, at present, is similar to that of other high-technology industries but well below that of the selected traditional industries.

Apparel	Furniture	Plastics	Industrial Controls	Sci. & Eng. Instruments
79.7	98.4	0.1	22.7	86.6
20.3	1.6	99.9	77.3	13.4
100.0	100.0	100.0	100.0	100.0
84.0	71.9	47.5	38.2	49.7
76.2	100.0	90.5	71.1	88.0
91.6	100.0	95.9	81.8	92.0
34.2	45.0	46.1	54.1	55.5
54.8	55.0	53.9	45.9	44.5
100.0	100.0	100.0	100.0	100.0
52.2	17.8	55.7	26.8	13.4
76.1	43.8	67.8	42.5	27.0
86.8	68.8	81.6	60.2	46.6

Microelectronics and Industrial Complexes

Analysis of the semiconductor industry in North Carolina as an industrial complex focuses on two questions. First, would new semiconductor manufacturers buy from and sell to particular North Carolina-based firms on a significant scale? In other words, can one expect important and special trade effects? Second, could other types of industrial agglomerations related to semiconductors occur, though not necessarily based on trade? As noted above, industrial complexes are sets of technologically linked industries that also locate in spatial proximity; the existence of complexes does not in itself assure that there is trade among the group of industries within the complex, but it does imply a strong potential for trade.[14]

Several studies have identified electronics, or communications–electronics, as an industrial complex in which SIC 3674 or 367 is a consistent member (see table 3).[15] The specific industrial composition of these complexes varies somewhat among the studies because of the

Table 2. Output Multipliers for Selected Industries in North Carolina

Industry No. (1972)	Industry	Output Multiplier
160100	Broadwoven fabric mills	2.66
180400	Apparel made from purchased material	2.64
220101	Wooden household furniture	2.38
280100	Plastics and resins	2.09
530500	Industrial controls	2.07
570200	Semiconductors and related devices	2.06
620100	Scientific and engineering instruments	2.02

Source: Tables of RIMS II Direct Coefficients, Total Multipliers, and Earnings Multipliers, State of North Carolina, 1982, produced from the RIMS II Regional Input-Output Modeling System, U.S. Dept. of Commerce, Bureau of Economic Analysis.

diverse research methods used, the different geographic areas surveyed, and the different techniques of data analysis employed. Because of uncertainty about which of these variations are most likely to develop in North Carolina, our analysis identifies the *potential* industrial complex in North Carolina as comprising *all* the industries listed in the relevant complex in each of these studies. The list thus represents the broadest potential membership of an industrial complex centered on the microelectronics industry in North Carolina, including the potential of enhanced interindustry trade within the state.

When this composite list of communications–electronics industries is compared to a list of industries from which the microelectronics industry purchases its inputs (backward-linked industries), only the electronic components industry features in both categories. The small degree of overlap suggests that spatial proximity to input-producing industries (aside from electronic components manufacturing, itself a purchaser of its own output) is unimportant to communications–electronics complexes. This is probably due to the relative unimportance of transportation costs and lack of need for face-to-face contact, as compared to other locational factors considered by the various input-producing industries and the electronic components industry. Indeed

Table 3. Composition of the Communications-Electronics Complex

SIC	Industry	Bergsman	Cherniack	Czamanski
357	Office machines & computers	x	x	
359	Machinery (not elsewhere classified)			x
361	Electrical distribution equipment	x		
362	Electrical industrial apparatus	x		
365	Radio & TV receiving equipment	x		x
366	Communication equipment	x		x
367	Electronic components	x	x	x
369	Misc. electrical machinery	x		
372	Aircraft & parts			x
381	Scientific instruments			x
382	Measuring & controlling instruments	x		
386	Photographic equipment	x		

Source: Joel Bergsman, Peter Greenston, and Robert Healy, "A Classification of Economic Activities Based on Locational Patterns," Journal of Urban Economics 2, no. 1 (1975); Howard Cherniack, "Clusters of Economic Activity: Submetropolitan Structures and Intermetropolitan Movement" (Ph.D. dissertation, University of North Carolina at Chapel Hill, 1980); Stan Czamanski, Study of Clustering of Industries (Halifax, N.S.: Institute of Public Affairs, Dalhousie University, 1974).

the principal location factors for industrial chemicals, nonferrous metals, and metal stamping products, the three most important inputs to the microelectronics industry after the electronic components industry itself, are likely to be quite different than those for the microelectronics industry. In particular, the geographically dispersed and sectorally diversified product markets of the input-producing industries, combined with large-scale production technologies, make spatial proximity to individual complexes even more unlikely. In addition, because many firms are multilocational, *interregional* flows of goods, services, and information are frequently more important than *intraregional* economic linkages.[16]

The industrial composition of the communications–electronics complex suggests that intraregional trade effects may be more likely to occur from forward linkages. Radio and television receiving equipment, communications equipment, computers, photographic equipment, measuring instruments, and medical instruments all purchase microelectronic devices *and* tend to locate in spatial proximity to electronic components manufacturing plants.[17] The principal question is whether this spatial proximity actually reflects forward linkages and hence enhancement of interindustry trade within the region, or whether it is due to some other types of agglomeration economies *not* based on trade. The argument against the former rests on the nature of the product markets of the microelectronics industry. Data from input-output tables and common knowledge both confirm that the immediate product market for microelectronic devices is highly diversified among user industries and is becoming more so. Furthermore, standardization of production has made possible large-scale, bulk production directed toward national and international markets. Increasing miniaturization has further diminished the importance of transportation costs. For these reasons, it is more likely that the observed spatial proximity of forward-linked industries to the microelectronics industry is due to the sharing of common factor inputs rather than to trade relations. The principal exceptions would be firms engaged in custom design and the production of highly specialized devices, such as equipment for research-based medical centers. This type of production would be small-batch and would require a relatively high degree of face-to-face contact between producer and customer.

Many of the industries comprising the communications–electronics complex are highly correlated in their locational patterns. These high correlations may be explained sufficiently by (1) a commonality in principal locational factors and (2) the ubiquity of such factors. Locational factors include accessibility to a highly skilled and technically trained segment of the labor force; close proximity to research-based universities and associated institutes; high "quality-of-life" attributes including environmental quality, residentially based amenities, and quality public education; and demonstrated state and local government policies and attitudes supportive of high-technology industry.[18] More fundamentally, a shared, distinctive type of milieu appears to be the glue that binds the communications–electronics complex of industries, which along with biomedical industries constitute a large proportion of the high-technology sector of the economy. The principal point is that this milieu need not be based on any level of interfirm activity such as trade or even the direct sharing of ideas or innovations. Rather it may be comprehended as a collective good produced and consumed as a result of common location, but nurtured by the activities of third-party institutions such as research-based universities, state-supported research centers (such as MCNC), and industry-wide sponsored activities.

These institutions would be supportive of another economic development impact that might be potentially generated from the growth of the microelectronics industry in North Carolina. Business services, such as software design, the design of management information systems, and other computer- and information-based producer services, may develop and locate in spatial proximity to the microelectronics industry. More generally, computer manufacturing may become more technologically linked to related services as computer software and other applications increasingly play a more important role in hardware design criteria. The need for high levels of face-to-face interaction between software applications designers and hardware designers would necessitate similar locational patterns. The critical question in assessing the likelihood that this type of economic development would occur is the extent to which formation of new companies might be expected. The potential for new business spin-offs from the growth of the microelectronics industry in North Carolina is considered below, after an analysis of the relative stability of microelectronics employment and the occupational structure of the microelectronics industry.

Microelectronics and Local Employment

The ability of an important local industry to withstand national cyclical fluctuations contributes to the overall socioeconomic well-being of a community. A cyclically stable employment base also enhances the fiscal strength of local and state governments through lower demands on unemployment insurance funds and other types of short-term relief, and through more efficient long-term utilization of public capital investments. Will the growth of the microelectronics industry increase, decrease, or not affect the stability of employment and income in North Carolina?

To approach this question we compared the recent cyclical performance of the microelectronics industry to that of other selected manufacturing industries, using national-level data. The percentage of change in industry employment levels from the peak of a U.S. business cycle to its trough (dated by turning points in the level of total nonagricultural employment, seasonally adjusted) over the four recessions

Table 4. Percentage of Change in Selected U.S. Industry Employment during Four post-1970 National Recessions, with Change for North Carolina Manufacturing Employment

	Mar 70 - Nov 70		Oct 74 - Apr 75		Mar 80 - Jul 80		Jul 81 - Sep 82	
Rank Order	Industry (SIC)	% Change	Industry (SIC)	% Change	Industry (SIC)	% Change	Industry (SIC)	% Change
1	367*	-16.2	367*	-18.8	251	-16.3	222	-14.7
2	381	-11.7	251	-18.2	25	-12.1	221	-14.0
3	DUR MFG	- 9.5	25	-17.8	228	-10.5	DUR MFG	- 9.8
4	36	- 8.5	3674**	-17.2	222	- 8.8	282	- 8.9
5	366	- 7.9	282	-15.2	DUR MFG	- 7.0	228	- 8.4
6	MFG	- 6.0	36	-15.0	36	- 5.3	MFG	- 7.3
7	357	- 5.6	228	-13.9	282	- 5.3	251	- 7.1
8	221	- 3.9	232	-13.4	MFG	- 5.1	232	- 6.2
9	28	- 2.9	221	-13.0	232	- 5.0	281	- 6.0
10	282	- 2.4	DUR MFG	-11.8	221	- 2.9	381	- 5.7
11	25	- 1.9	MFG	-11.0	NON DUR MFG	- 2.1	25	- 4.5
12	NON DUR MFG	- 1.0	NON DUR MFG	-10.0	367*	- 1.6	28	- 4.5
13	232	- 0.1	28	- 7.9	28	- 0.6	36	- 4.5
14	222	0.0	381	- 7.5	281	- 0.6	NON DUR MFG	- 3.1
15	228	1.5	222	- 6.4	366	0.5	366	0.6
16			357	- 6.2	381	1.7	367*	2.2
17			366	- 4.5	3674**	2.1	3674**	3.4
18			281	- 1.9	357	3.2	357	5.1
NC MFG		- 2.2		-12.7		- 3.3		- 5.5

Source: U.S. Department of Labor, Bureau of Labor Statistics, Employment and Earnings, monthly issues in 1970, 1974, 1975, 1980, 1981, 1982; North Carolina Employment Security Commission, North Carolina Labor Force Estimates, 1970, 1974, 1975, 1980, 1981, 1982, and State Labor Summary (monthly).

since 1970 served as a measure of the ability of an industry to withstand the deleterious effects of national cyclical fluctuations (see table 4). The data indicate that the microelectronics industry's cyclical performance was well below the average for manufacturing as a whole in the 1970–71 and 1974–75 recessions and, in fact, poorer in *both* periods than every comparative industry except furniture. In the 1980–81 and 1981–82 recessions, however, the microelectronics industry outperformed every industry except SIC 357 (electronic office equipment and computers), with positive gains in employment levels in both periods.

Comparison of the cyclical performance of the U.S. microelectronics industry to manufacturing as a whole in North Carolina over the same periods presents much the same picture (table 4, lowest row). Though it fared comparatively worse in the two earlier recessions, the U.S. microelectronics industry outperformed the North Carolina manufacturing sector in both the 1980–81 and 1981–82 recessions. The actual effect of microelectronics industry development on the stability of the state's economy will depend, in part, on the interaction between the industry's cyclical behavior and the industrial structure of the state. If interindustry relations between microelectronics and other industries in the state are relatively weak, however, that effect may be negligible.

The microelectronics industry has been neither consistently stable nor unstable over national business cycles. Since 1970 the industry has displayed notably different cyclical performance from that of the selected comparative industries and that of manufacturing as a whole. This idiosyncratic behavior is a reflection of the different circumstances behind, and causes of, the four recessions. If the nature of national business cycles in the remainder of the 1980s proves similar to that of the period from 1980 to 1982, growth in the microelectronics industry should contribute to the stability of North Carolina's employment and income base.

Occupational Structure

Although the job-generating potential of the microelectronics industry in North Carolina has received much attention, less has been said about the *types* of jobs that would be created. The occupational structure of an incoming industry determines, in large part, whether average wage and salary levels in local economies will be raised or lowered. Along with knowledge of other components of local labor supply, the particular occupational structure that results from industrial growth is the key to determining whether unemployment rates will go down, the level of demands placed on education and training programs to meet labor skill requirements, and the likely extent and type of labor and population in-migration to the state.

The occupational structure of an industry varies *among* establishments in the industry, because of functional specialization in different establishments and also because of spatial variations in technology. We have used national data to compare the average occupational structure of the microelectronics industry to that of other selected manufacturing industries. The actual occupational structure of the microelectronics industry in North Carolina may vary considerably from the industry average in certain labor market areas.

Table 5 shows the occupational distribution of SIC 367 in comparison to selected three-digit U.S. manufacturing industries and two-digit North Carolina industries. First, it is evident that the occupational distribution of SIC 367 differs considerably from the nondurable goods industries concentrated traditionally in North Carolina (SIC 221, 228, 231, 251). There is a higher proportion of managerial, professional, and technical workers and about an equally lower proportion of production workers in SIC 367 compared to these "traditional" industries (which among themselves have nearly identical distributions for the major occupational groupings). Compared to other selected, emerging growth industries in North Carolina, however, the microelectronics industry has a lower percentage of the most highly skilled occupational groups. Indeed communications (SIC 366), scientific instruments (SIC 381), and industrial inorganic chemicals (SIC 281) each have distinctly higher proportions of professional, technical, and managerial workers and distinctly lower proportions of production workers than SIC 367. A more detailed analysis of the skill levels of workers in the microelectronics industry (see Sampson, chapter 10, this volume) indicates that less than 10 percent of the production and maintenance work force is classified in skilled occupations.

If the occupational structure of the emerging microelectronics industry in North Carolina were to reflect that of the U.S. average for SIC 367, the skill demands placed on local labor markets would be considerably higher than levels expected by the traditional North Carolina industries, with resulting pressure placed on the state's education and job training programs. On the other hand, other high-technology industries that may be emerging in the state in the 1980s have larger proportions of professional and technical workers and lower proportions of production workers than SIC 367. The comparatively highly stratified work force of the microelectronics industry—very highly skilled engineers, scientists, and technicians concentrated at the top and many semiskilled and unskilled assembly workers employed for production, with a relatively small proportion of middle-level, skilled occupations—would here again place very different demands on local labor supply. The specific structure of occupational demand will depend upon which segment of microelectronics manufacturing actually locates in North Carolina (see below). The staffing patterns of those establishments principally will determine whether the occupational demand structure will resemble that for other emerging high-technology industries (such as SIC 366, 381, and 281) or approximate the occupational structure of the more traditional industries in North Carolina (SIC 22, 23, and 25).

Local Labor Market Impacts

A complete assessment of the impact of growth in the microelectronics industry on the local labor market in North Carolina, which would take into account both demand and supply, is beyond the scope of this chapter. Instead we have focused on the likely variation in the pattern of labor demand across North Carolina, with implications for the local wage structure, labor supply requirements, unemployment rates, and net labor and population migration in different types of local labor markets.

Aside from the scale of growth in the state as a whole, the nature of impacts on the local labor market will depend on the spatial distribution of the different phases of the manufacturing process. Because growth in the microelectronics industry will not occur uniformly across regions in the state, the scale effects will be important in areas where growth clusters. The likely spatial pattern of such growth will reflect the attractiveness of various labor markets around the state to each of the three phases of the manufacturing process: product design, wafer fabrication, and assembly.[19]

The first phase, product design, encompasses a high degree of research and development activity and thus involves a high proportion of highly skilled professional and technical workers. Proximity to universities and other nonproprietary research and development activities, proximity to a well-served airport, and high levels of residential amenities and public services to attract and retain this type of labor force are the primary factors that firms consider when choosing locations for such facilities. There seems to be little question that this phase of the manufacturing process would locate in Wake, Durham, and Orange counties with the Research Triangle Park as the focus.

The second phase, wafer fabrication, is advanced manufacturing. The

Table 5. Occupational Staffing Patterns for Selected Manufacturing Industries, U.S. and North Carolina, 1980

U.S. Occupational Category	Industry (SIC)						
	20	22	221	222	268	25	251
Managers and offices	6.4	3.7	2.7	2.8	2.6	5.3	4.7
Professional	2.7	1.7	1.8	1.9	1.2	2.2	1.6
Technical	0.7	0.4	0.9	0.8	0.6	0.8	0.4
Service	3.1	2.2	2.7	2.3	2.5	1.9	2.0
Production	73.2	81.9	85.2	84.6	86.5	76.4	79.4
Clerical	10.0	8.8	6.6	7.1	6.4	11.1	9.7
Sales	4.0	0.8	0.2	0.4	0.2	2.3	2.2
North Carolina	**20**	**22 (except 221)**				**25 (except 2**	
Managers and offices	5.7	2.9				3.7	
Professional and technical	1.8	2.0				1.9	
Service	2.9	2.1				2.5	
Production	77.4	83.4				81.2	
Clerical	8.0	8.3				8.4	
Sales	4.0	0.6				1.7	

Source: U.S. Dept. of Labor, Bureau of Labor Statistics, Occupationa Employment in Manufacturing Industries, Bulletin 2133 (Washington, D.C., September 1982); Employment Security Commission of North Carolina, Labor Market Information Division, Occupational Employment in Manufacturing and Hospitals, North Carolina 1980 (Raleigh, 1982).

labor requirements include a mixture of highly skilled professionals (such as engineers), technicians able to operate sophisticated instruments and equipment, and semiskilled workers for the more routine tasks. The professional work force involved is proportionally smaller than in the product design phase, but greater than in the assembly phase. In this case the principal factor in locational decisions is ac-

Industry (SIC)									ALL MFG
28	281	282	36	362	366	367	38	381	
9.9	9.3	6.8	6.4	5.4	8.2	6.6	8.7	9.2	6.6
12.3	13.7	10.1	12.3	8.4	22.5	11.5	11.6	17.5	6.9
5.3	5.8	5.2	6.2	4.8	9.8	7.4	7.4	10.0	2.9
2.2	2.6	1.6	1.4	1.4	1.1	1.3	1.7	1.5	1.8
52.6	55.9	65.4	60.3	68.2	41.3	61.0	53.7	42.9	68.1
14.6	11.6	9.7	12.3	10.8	16.3	11.1	14.1	17.3	11.5
3.2	1.0	1.1	1.0	1.0	0.6	1.0	2.8	1.7	2.2

28	36 (except 361)	38 (except 387)
5.0	3.6	5.0
10.6	11.9	8.8
1.5	0.9	1.8
71.9	73.9	71.8
8.7	9.3	10.8
2.2	0.5	2.0

cess to a highly skilled and technical labor force but not necessarily to universities and research enterprises. In North Carolina, wafer fabrication plants would probably locate within or immediately adjacent to major metropolitan regions in the Piedmont, including Raleigh–Durham, Greensboro–Winston-Salem–High Point, and Charlotte–Gastonia.

The third phase of manufacturing, known as assembly, or bonding, is highly labor-intensive and routinized. Labor costs and the "labor climate" are the primary factors in locational decisions for such plants. For this reason assembly operations have increasingly tended to locate in the less developed countries (the so-called offshore platforms). North Carolina's attraction here would be based chiefly on the state's relatively low wage structure and its largely nonunionized and lower-skilled labor force. Within the state, assembly operations would be likely to locate in nonmetropolitan regions that have the lowest wage levels, an adequate supply of less skilled labor, and good transportation with close proximity to the interstate highway system. Even so, it is unlikely that a large number of assembly operations would locate in the state despite these advantages relative to other regions of the United States. Unless worldwide economic and political changes occur that would reduce U.S. direct foreign investment, the trend toward overseas location should continue.

This review of the differential locational behavior of the three manufacturing phases of the microelectronics industry provides strong clues as to the *directions* that local labor market impacts may take in North Carolina. If product design facilities locate in the Triangle counties, growth in the microelectronics industry should increase that area's per capita and per family income levels, because the wage and salary structure in the product design phase of microelectronics is higher than the projected area average. As a result of increased competition among existing establishments for scarce labor skills in the professional, technical, and some clerical occupations, upward pressure will be felt on wages and salaries in other sectors of the local labor market; this is often termed a "wage rollout." Increases in labor supply from existing area sources, including the current unemployed and underemployed, those enrolled in educational and training programs, and highly skilled workers in other industries, will probably not match the labor requirements for even moderate growth in the microelectronics industry. As a result, there may be significant in-migration of workers for certain jobs, especially in the most highly skilled, highest-paid occupations. The net result would be little effect on the area's already low projected unemployment rate, but upward pressure on prices, particularly for housing. The relative capacity of the state's community college system to respond to the labor skills demanded will affect the rate of in-migration, the amount of upward pressure on wages and salaries in the area's existing industries, and the rate of growth of the microelectronics industry itself.

Effects of the location of wafer fabrication plants in or immediately adjacent to the metropolitan areas in the Piedmont region of the state are more ambiguous. Metropolitan per capita and per family income would probably rise. The effect on area unemployment rates would depend largely on the success of job training and public education efforts targeted at the disadvantaged and on firms' willingness to cooperate in such efforts. It is likely that a substantial number of professionals would be brought to these areas from outside the state; the degree of in-migration of skilled production workers would again depend upon the capacity of the state's various education and training programs to supply such demands from within the local labor markets. Upward pressure on wages in existing industries due to labor scarcity would be felt more in the skilled production occupations than in the professional occupations. The extent of this wage rollout would depend on the size of the wage differential between skilled production workers in the microelectronics industry and others in the area's traditional industries that utilize such labor. Unionized microelectronics firms, although atypical of the industry, would create the largest rollout effect on the local labor market.

The location of assembly, or bonding, plants in predominantly nonmetropolitan areas of the state would have the most dramatic impacts, if only because of their relatively large-scale effects in small labor markets. The high proportion of women employed in microelectronics assembly work nationally suggests that women's participation in the local labor force would increase (see Sampson, chapter 10, this volume). Average family income should increase as a result. However, if married female workers were predominantly hired instead of unemployed or underemployed primary (male or female) workers, inequality in family income might increase. Thus the likely impact on local unemployment rates remains ambiguous. Some amount of in-migration from areas in the state with higher rates of unemployment would be expected. Wage levels would be tied to the local wage structure, so there should be little or no effect on the area's average wage level. The key factors in predicting local labor market impacts would be firms' hiring practices, changes in female labor force participation, and the level of area unemployment, rather than presence of education and training programs as in the cases discussed above.

Fiscal Impacts

For state and local governments the fiscal impacts of growth in microelectronics hinge on the considerable uncertainty surrounding the industry's future, especially in the application of new technologies. Although the qualitative differences reported below are not highly significant, the risk of large positive or negative effects would appear to be greater in microelectronics than in comparable high-technology industries. Given this higher risk, government jurisdictions would be wise to monitor the microelectronics sector regularly and develop contingency plans to deal with possible fiscal threats and opportunities.[20]

In recent years, fiscal impact studies have been widely applied. Unfortunately, in most cases revenues and expenditures are analyzed only for broad categories of development—for example, industrial, commercial, or residential projects. Systematic data on interindustry differences within the manufacturing sector are not available. Moreover, the aggregate figures that are available give approximations of average revenues and costs per employee, providing no insight into marginal differences between development projects.

Most of the industrial demand for public services falls on public works (roads, sanitation, sewerage, and water supply) and public safety (police and fire protection).[21] Compared to other manufacturers, new microelectronics facilities are unlikely to have an extreme impact on land use and related aspects of transportation. In fact, the effectiveness of local growth management should be a greater determinant of the degree of impact than the size or employment characteristics of microelectronics facilities. It is safe to say that local planners will realize the considerable importance of quality of life issues to microelectronics concerns and will be more likely to propose and execute effective growth management strategies as a result. Thus the overall impacts of growth in microelectronics should be no more severe than for comparison industries and may well be more beneficial. Demands on water supply and waste management would be less than with some other high-technology industries.[22]

As for public safety, microelectronics facilities should not place greater demands on local police services than other manufacturers. The same security issues will pertain. However, it should be noted that industrial espionage has become a problem in Silicon Valley. There have been instances of illegal trading of industrial secrets and a black market for wafers and chips.[23] The replication of such activities in North Carolina may unduly burden state law enforcement agencies and strain limited state personnel that are already involved in major law enforcement efforts. The relative demand for fire protection services may be gauged by comparing the insurance ratings of various manufacturing concerns, in terms of property insurance coverage. For comparable structures, the ratings for semiconductors, industrial controls, and engineering and scientific instruments are about the same. Ratings for textiles and apparel are slightly higher, followed by plastics and, finally, furniture. Thus it would appear that semiconductors would create less demand for fire protection than traditional industries and no more than other emerging high-technology industries.

Microelectronics businesses will pay taxes to state and local governments on real property, inventory, and income. The semiconductor industry would pay an average amount of property taxes compared to traditional and emerging industries in North Carolina. This conclusion is based on a comparison of the book value of depreciable assets per dollar of value added for eight four-digit industries in 1977. Semiconductors had eighty cents' worth of capital stock in place per value-added dollar, about the average amount for the eight industries.[24] Similarly, the semiconductor industry would pay about the average amount of taxes on its inventory. If the 1977 pattern holds in the future, the ratio of inventory to total assets will place the semiconductor industry in the middle of the distribution of comparison industries; in 1977 four had higher ratios, three had lower ratios. The semiconductor industry would likewise pay about the average amount of taxes on earned income. Earned income was calculated by subtracting total payrolls and depreciation charges from value added and dividing this figure by total assets to derive a ratio of earned income per dollar of invested capital. Again, the industry's ratio was near the middle of the 1977 distribution. There appears to be nothing extreme about the public revenue-generating potential of microelectronics business.

Growth in the microelectronics industry also means additional public expenditures and revenues generated by employees and their families. Because available studies estimate fiscal impacts for the "average" employee in manufacturing, they do not provide sufficient information to pursue this issue very far. Adequate analysis would require demand profiles for public services and supply profiles for public revenues that permitted a comparison of microelectronics employees with employees of other concerns. The closest approximation that can be made is a comparison of the occupational profiles of different industries, assuming that each industry's unique profile indicates a different pattern of fiscal impacts. As seen in table 5, microelectronics and other emerging industries have similar staffing profiles; their employees should thus generate similar demands on public goods and services. That is, growth in the microelectronics industry should not generate unique public expenditures through its employees. Because different occupational profiles imply different patterns of income distribution, and hence of consumption and housing, one can assume they also indicate differences in generation of state and local revenues through taxes on income, sales, and property. Table 5 thus suggests that microelectronics and other emerging industries share similar potential for generation of revenues through employees; on average such households would produce more revenue per capita than households of employees in traditional industries in North Carolina.

In summary, the qualitative fiscal impacts of growth in microelectronics should be about the same as for other emerging industries and slightly better than for traditional industries. State and local governments should be able to accommodate this growth without difficulty, despite the greater uncertainty that surrounds the industry's future.

Formation of New Companies

One of the most appealing features of growth in microelectronics is its legendary capacity to breed new businesses, not only through new applications of high-technology products, but also through "spin-offs" within the industry itself. If North Carolina is successful in attracting microelectronics concerns during the 1980s, what type of firms will be represented? Two factors must be considered to answer this question knowledgeably. First, changes in the structure of the microelectronics industry as a whole must be gauged. Second, the segment(s) of the industry most likely to be attracted to the state must be identified.

Semiconductor production grew at a compound annual rate of 16.2 percent over the decade from 1972 to 1981. Industrial output in constant of 1972 dollars exceeded $11 billion per year in the early 1980s. Industrial growth through 1986 is estimated to be above 11 percent per annum. Both merchandise exports and imports of semiconductors are assuming greater importance in the industry, with imports growing at a slightly faster rate.[25]

Two emerging trends should be highlighted: concentration and consolidation. The proliferation of new firms, a dominant feature of the industry prior to 1980, is unlikely to continue in the near term, for several reasons. First, the 1981–82 recession frustrated industry sales projections, generating excess capacity and, for the first time since 1974, personnel layoffs on a substantial scale.[26] Second, U.S. producers are facing stronger foreign competition for market share, especially from Japanese firms. Third, 64K RAM applications require much heavier financial commitments ($50 million to $100 million per development project) than previous state-of-the-art technologies. Increasing emphasis on computer-aided design (CAD) and manufacturing (CAM) also favors larger business entities with sufficient working capital to purchase the necessary capital equipment. Thus microelectronics has begun to experience the "shakeout" of marginal firms that typically occurs as an industry matures, although rapid technological change and new product development have delayed the process. Wafer fabrication, assembly, and product design should become more concentrated in larger firms. Acquisitions, mergers, business failures, and withdrawals of certain companies from this market—all of which promote concentration—will occur with increasing frequency. However, there will remain considerable room for formation of new companies in market segments devoted to product design, software applications, and customized products.

Consolidation has increased in connection with the vertical integration of the electronic computing equipment industry. Most large computer manufacturing firms control research and development, prototype design, wafer fabrication, assembly, and distribution through semiconductor subsidiaries. These subsidiaries are strengthening intrafirm linkages by coordinating research and development, semiconductor production, and computer production more closely. Over the past few years, however, consolidation has slowed as the mixed results of earlier acquisitions have become known. U.S. computer manufacturers now seem committed to supporting a healthy and autonomous domestic semiconductor industry. Many computer firms are entering into joint ventures or making equity infusions in semiconductor firms with this objective in mind.[27]

North Carolina is not likely to attract a random sample of microelectronics concerns that emerge from the concentration and consolidation in the industry. The composition of the microelectronics industry in the state may well depend upon the focus of recruitment efforts by MCNC, whose leadership has considerable latitude of choice. The industry's composition will largely determine the potential for formation of new companies in the state. Under one scenario, the typical microelectronics establishment recruited to North Carolina would be a branch plant or subsidiary of a large computer or consumer electronics firm. As such, these establishments would mirror the characteristics of most companies already located in the Research Triangle Park, tending to be research subsidiaries or branch plants of larger corporations headquartered in the East. Under another scenario, the state could recruit young, independent microelectronics firms and the related software or supportive service firms. Silicon Valley has amply demonstrated that this segment of the microelectronics industry has great potential for generating new companies.[28] Which scenario is more likely for North Carolina? This question may be addressed by examining the relations between the industry, the relevant economic actors, and the potential locations for the industry within the state.

It is important to emphasize that the economic actors initiating successful new enterprises are usually experienced corporate executives. New companies are not likely to be initiated by production workers, their supervisors, technicians, scientists, or even plant managers of semiconductor facilities, regardless of their various talents and skills. Many of the most prominent microelectronics firms, such as Tandem and Apple, were started by former Hewlett-Packard executives.[29] Without direct managerial experience in the industry at a reasonably high level of responsibility, experience in the planning and execution of projects, or experience as a new-venture manager, the probability of starting a successful new enterprise is very low.

This assessment may appear too narrow and restrictive. Surely, new companies can be formed by other actors such as the applied scientists who would be engaged in microelectronics research at North Carolina universities and other research facilities. The computer software firm SAS, Inc., spawned by computer scientists from North Carolina State University several years ago, is a case in point. However, as a general rule, very few successful new companies have evolved from the diverse array of research activities that have been conducted in the Research Triangle Park. Indeed, it may be possible to test the validity of this argument by monitoring the medical and biological research sectors in the Research Triangle area, where the critical density of skilled researchers and managerial talent that breeds business spin-offs may well be already in place. It will be instructive to see whether successful businesses can be developed by the scientists or administrators involved in such work.

One popular view is that formation of new companies can be achieved through special efforts to provide entrepreneurship training, management assistance, and start-up capital. The probability of success through such efforts alone is not very high. The most relevant proximate causes of forming a new company have more to do with threats and opportunities perceived by actors with considerable business experience. On the positive side, executives must be able to perceive profitable opportunities in new products, markets, processes, or organizational forms and to crystallize their vision in viable business plans. On the negative side, executives must experience some form of "displacement" pressure, feeling the need to make a change. Displacement may be caused by top management's rejection of the executive's ideas or the sense that the executive's future with the company is growing uncertain. In other cases, displacement pressure may come from personal sources, often related to life cycle changes or to reaching age plateaus (for example, turning forty or fifty). Usually it is a combination of such positive and negative forces that propels an executive toward entrepreneurial activities.[30]

Besides the characteristics of the actors themselves, it is important to examine relevant features of labor markets within the state that are likely to receive microelectronics facilities. Under the first scenario suggested above, large facilities settling in the state would center their research and development and product design activities in the Research Triangle area and locate related wafer fabrication in or adjacent to the largest metropolitan areas. Under the second scenario, executives of small, independent firms would seek similar locations that offered high quality of life conditions, proximity to local suppliers, and good access to national markets. However, these new firms might not be able to afford the prime sites within the larger metropolitan areas.

Nor is it clear that these metropolitan areas would be able to support the formation of new companies in high-technology industries that serve national markets. Although research on this point has not been complete or decisive, it does seem that a particular "milieu" is required for the adoption and promotion of entrepreneurship.[31] This milieu is probably more complex and specialized than most descriptions of urbanization economies or urban amenities would suggest. Surely more is involved than the presence of basic economic factors, such as access to markets or availability of inputs, that are found in all areas of the state. Although North Carolina has one of the most favorable business climates in the nation,[32] that particular business climate may not be supportive of entrepreneurship. The largest metropolitan areas, which are developing an infrastructure of financial services, public facilities, university and professional education, skilled workers and technicians, and business services, may be able to accommodate entrepreneurial activity.

Anecdotal evidence suggests that quality of life considerations are important to entrepreneurs. If they enjoy an area, they are more likely to remain there while they are trying to initiate successful ventures. Most start-ups fail, but entrepreneurs learn from their mistakes and try again about three years later.[33] If such individuals choose to remain in North Carolina as they gain business experience and learn from past mistakes, they could make a valuable contribution to new employment and economic diversity in their second or third attempts. Unfortunately, quality of life is not a factor that can be measured objectively. It essentially depends upon the perceptions of the entrepreneurs in question. One study of entrepreneurship in Minnesota, however, does conclude that continued investments in social and physical infrastructure will enhance a state's attractiveness to entrepreneurs.[34]

Because venture capital has been a significant source of financing for businesses in high-technology sectors, the lack of local venture capital may appear to be an important barrier to the formation of new microelectronics companies in North Carolina. But several important facts suggest otherwise. The venture capital industry is an aggressive and growing industry that is organized to serve a national market. Venture capitalists have demonstrated their ability to find attractive investment opportunities wherever they may be. For example, venture funds managed in the Boston area and in other major money centers have financed many new microelectronics firms initiated in California, Texas, and elsewhere. Furthermore, public and private venture capital sources are growing in North Carolina. The state also has an active, competitive commercial banking and thrift industry. If good business ideas emanate from the local microelectronics industry, sufficient venture capital should be available to bring these ideas to the marketplace.

Thus it would appear that the most critical factors affecting the formation of new microelectronics companies in North Carolina will be the type of firms that locate in the state and the character of their executives. Under the first projected scenario, in which microelectronics facilities would primarily be subsidiaries or branch plants of large corporate entities, the most talented managers would eventually leave the state to assume positions of greater responsibility at headquarters or at other company locations. Under the second scenario, in which independent microelectronics firms would predominate, gradually some of the most talented executives would form new companies on their own, many choosing to remain in the state. A crucial issue is whether innovative ideas coming out of MCNC, area universities, and local microelectronics firms are absorbed by in-state branches of national corporate entities and subsequently applied elsewhere, or whether these ideas are absorbed and applied locally by North Carolina-based firms.

Summary and Conclusions

It is difficult to forecast the economic development effects of an industry whose technology is still rapidly changing and which is just beginning to locate in the region. Discussion above has relied on a variety of systematic, analogical, and qualitative techniques for analysis and forecasting. The best available evidence suggests that the growth of the microelectronics industry in North Carolina will lead to a pattern of economic development impacts different from those of the traditional manufacturing industries in the state. They will not be unique, however, in comparison to impacts of other emergent manufacturing industries in the state, such as drugs, plastics, nonelectrical machinery, and instruments. State and local multiplier effects will be positive but not dramatic. Interindustry trade linkages may be weak, because firms in this industry tend to purchase material inputs from suppliers serving national markets that are generally located outside the state. The effect on average earnings should be positive because of the probable tight labor market in the Triangle area.

There is greater uncertainty about the amount of induced economic development due to the location of the microelectronics industry in the state. Particular milieux, characterized by a concentration of research and development activity, research-oriented universities and medical centers, and a set of public services and facilities demanded by highly educated professionals, should be attractive to firms in the business service industries, including software design, management consulting, and a variety of other information-based or computer-based industries.

Impacts on employment and labor markets will depend on the degree to which the different phases of the production process locate in the state. It is likely that the product design phase would be attracted to the Triangle area, whereas the wafer fabrication phase would locate in, or immediately adjacent to, the three largest metropolitan areas in the state: Raleigh–Durham, Greensboro–Winston-Salem–High Point, and Charlotte–Gastonia. Unless relative labor costs and political conditions change enough internationally to alter the industry's preference for foreign locations, it is not likely that a large amount of the assembly or bonding phase of the industry will locate in North Carolina. Thus the economic benefits of growth in microelectronics will be concentrated in the metropolitan areas of the state, where unemployment problems are least severe. If shortages of certain types of professional and technical labor develop in these areas, wage and salary levels for these occupations could rise as existing businesses increase their wage and salary offers to remain competitive in the local labor market. Effects on wage levels, in-migration of some types of skilled workers, and general labor supply for the microelectronics industry and other high technology industries will partly depend on the response of the community college system and other skill-training institutions in the state.

The fiscal impacts of growth in the microelectronics industry are uncertain because the technology is changing and good empirical studies are scarce. Given the available evidence, however, there is no reason to assume that the demand for public services generated by either the industry or its employees will be excessive. Projected impacts of the microelectronics industry on tax revenues are about average compared to other manufacturing industries, standardizing for scale of facility.

The potential business spin-offs from growth in the microelectronics industry could be minimal or substantial. One scenario discussed above is that microelectronics facilities locating in the state will be branch plants of national corporations, with little potential to spawn new companies. A second scenario is that independent microelectronics firms eventually will come to dominate the industry in the state, with a much greater potential for business spin-offs. External economic pressures and the maturing of the industry as a whole will continue to affect business location decisions, but state policy that encourages industrial recruitment and formation of new companies will have strong influences on the development and vitality of the microelectronics industry in North Carolina.

Notes

1. "How to Get the Economy Growing Again: The Search for a New Policy," *Business Week*, 8 November 1982, 108–9.
2. "Is Talk of High Tech Jobs More Political Than Real?" *New York Times*, 24 October 1981, E3.
3. Edward J. Malecki, "Public and Private Sector Interrelationships, Technological Change, and Regional Development," *Papers of the Regional Science Association* 47 (1981): 121–37.
4. "The New Faith in High Tech," *Wall Street Journal*, 27 October 1982, 32.
5. John S. Hekman, "What Are Businesses Looking For? Survey of Location Decisions in the South," *Economic Review* 67 (June 1982): 6–19.
6. See Norman J. Glickman, *Econometric Analysis of Regional Systems* (New York: Academic Press, 1977); and Saul Pleeter, ed., *Economic Impact Analysis: Methodology and Applications* (Boston: Martinus Nijhoff, 1980).

7. United National Industrial Development Organization, *Guidelines for Project Evaluation* (New York, 1972).

8. Price effects are gauged as changes in relative commodity prices, including rents and wage rates; quantity effects, as changes in physical inputs and outputs; income effects, as changes in the value of production, revenues, personal income, and so forth; wealth effects, as changes in the value of local assets as realized through capital gains and losses.

9. See Hollis Chenery and T. Watanabe, "International Comparisons of the Structure of Production," *Econometrica* 26 (1958): 487–521; Albert Hirschman, *The Strategy of Economic Development* (New Haven: Yale University Press, 1958).

10. See especially Stan Czamanski, *Study of Clustering of Industries* (Halifax, N.S.: Institute of Public Affairs, Dalhousie University, 1974); Stan Czamanski and D. Z. Czamanski, *Study of Formation of Spatial Complexes* (Halifax, N.S.: Dalhousie University, 1976); Stan Czamanski and Luiz Augusto Ablas, "Identification of Industrial Clusters and Complexes: A Comparison of Methods and Findings," *Urban Studies* 16 (1979): 61–80. Studies designed to identify industrial complexes have also been done by Joel Bergsman, Peter Greenston, and Robert Healy, "A Classification of Economic Activities Based on Locational Patterns," *Journal of Urban Economics* 2, no. 1 (1975): 1–28; and Howard Cherniack, "Clusters of Economic Activity: Submetropolitan Structures and Intermetropolitan Movement" (Ph.D. dissertation, University of North Carolina at Chapel Hill, 1980).

11. Figures are from Dun's *Census of American Business* (New York: Dun & Bradstreet, 1982).

12. The published data on interindustry linkages are rather old. The most complete information is contained in U.S. Department of Commerce, Bureau of Economic Analysis, *The Input-Output Structure of the U.S. Economy, 1972* (Washington, D.C., 1979). The basic data used to construct the 1977 table are in U.S. Department of Commerce, Bureau of the Census, *1977 Census of Manufactures* (Washington D.C., 1981), but the table itself did not appear until late in 1984. More current information can be found in recent volumes of U.S. Department of Commerce, Bureau of Industrial Economics, *U.S. Industrial Outlook*. For this study, industry experts in the Bureau of Economic Research and the Electronic Industries Association were also consulted.

13. U.S. Department of Commerce, Bureau of Economic Analysis, *RIMS II Regional Input-Output Modeling System* (Washington, D.C., 1981).

14. Only raw data on the origin and destination of inputs and outputs by industry provides prima facie evidence of trade relations. Since such data are very expensive to collect and assemble, researchers frequently resort to indirect measures of intraregional interindustry trade.

15. Bergsman, Greenston, and Healy, "Classification of Economic Activities," Cherniack, "Clusters of Economic Activity," and Czamanski, *Study of Clustering*, have all identified industrial complexes in U.S. metropolitan economies. A review of the analytic techniques and data used in these studies

is found in Czamanski and Ablas, "Identification of Industrial Clusters and Complexes."

16. Allan R. Pred, "The Interurban Transmission of Growth in Advanced Countries: Empirical Findings versus Regional Planning Assumptions," *Regional Studies* 14 (1976): 419–25; idem, *City Systems in Advanced Economies* (New York: John Wiley & Sons, 1977); John Rees, "Regional Industrial Shifts in the U.S. and the Internal Generation of Manufacturing in Growth Centers in the Southwest," in *Interregional Movements and Regional Growth*, ed. W. C. Wheaton (Washington, D.C.: The Urban Institute, 1981).

17. See, for example, Department of Commerce, Bureau of Economic Analysis, *The Input-Output Structure of the U.S. Economy, 1972*; and Czamanski and Ablas, "Identification of Industrial Clusters and Complexes."

18. U.S. Congress, Joint Economic Committee, *Location of High Technology Firms and Regional Economic Development* (Washington, D.C., 1982); see also Malizia, chapter 7, this volume.

19. Annalee Saxenian, "The Urban Contradictions of Silicon Valley: Regional Growth and the Restructuring of the Semiconductor Industry" (Working paper, Department of City and Regional Planning, University of California, Berkeley, 1981).

20. Emil Malizia, "Contingency Planning for Local Economic Development," *Environment and Planning B* 9 (1982): 163–76.

21. See R. W. Burchell and David Listokin, *The Fiscal Impact Guidebook: Estimating Local Costs and Revenues of Land Development* (Washington, D.C.: Department of Housing and Urban Development, 1979), 157.

22. Electronic components manufacturing (SIC 367) requires less water and produces less industrial waste than the inorganic chemicals industry (SIC 286), office and computing machines (SIC 357), and engineering and scientific instruments (SIC 381). See Department of Commerce, Bureau of the Census, *1977 Census of Manufactures*.

23. Susan Benner, "Storm Clouds over Silicon Valley," *INC.*, September 1982, 84–94.

24. Discussion throughout this paragraph is based on data from Department of Commerce, Bureau of the Census, *1977 Census of Manufactures*.

25. *U.S. Industrial Outlook, 1981*, 236–37.

26. In 1981, both total national employment and employment of production workers dropped by 2.2 percent. See *U.S. Industrial Outlook, 1981*, 238.

27. "IBM and Intel Link Up to Fend Off Japan," *Business Week*, 10 January 1983, 96–98.

28. The dichotomy between national corporations and independent firms should not be taken too literally. Some large corporations have spawned many entrepreneurs, just as some small firms are staid. It would be more accurate to group business entities according to their organizational cultures with respect to their capacity to encourage and absorb entrepreneurial activity.

29. "Can John Young Redesign Hewlett-Packard?" *Business Week*, 6 December 1982, 75.

30. Albert Shapero, *The Role of Entrepreneurship in Economic Development at the Less-than-National Level* (Washington, D.C.: U.S. Department of Commerce, Economic Development Administration, 1977).
31. Shapero, *The Role of Entrepreneurship.*
32. Alexander Grant & Co., *A Study of Manufacturing Business Climates in the 48 Contiguous States of the U.S.*, Center for Policy Studies, National Attitudes and Awareness Study (Washington, D.C., 1980).
33. Shapero, *The Role of Entrepreneurship.*
34. John Borchert, "Entrepreneurship and Future Employment in Minnesota" (Paper prepared for the Commission on Minnesota's Future, 1975).

10 Employment and Earnings in the Semiconductor Electronics Industry: Implications for North Carolina

Gregory B. Sampson

Recent studies of the impact of microelectronics and microelectronics-based technologies on modern industrial societies have highlighted their employment effects.[1] However, because most of these new technologies are just a little over a decade old in terms of widespread applications, most students of microelectronics and its social impacts have had to confine their remarks to very broad generalizations about the overall effect of the "micro-revolution" on employment. Even today there are only a few good case studies of the specific effects of these new technologies on employment levels and characteristics. Seemingly, these few empirical studies have only served to sharpen the debate over the direction, character, and scope of employment changes due to the development and introduction of these new technologies.[2]

The purpose of this chapter is to examine employment and earnings in the semiconductor electronics industry and their implications for the recent efforts in North Carolina to make the state a new center for the microelectronics industry. Much of the recent debate, however, centers not on the projected growth in or character of employment in the semiconductor industry itself, but rather on the employment effects due to the pervasiveness of microelectronics technology. A major concern is the ongoing or anticipated incorporation of integrated circuits and microprocessors into an almost numbing array of manufacturing production processes and industrial and consumer goods and the expected changes in levels and characteristics of employment across virtually the entire industrial structure of modern nations.

In broad outline the opposing schools of thought represent a continuation of the debate over the expected impacts of automation that broke out during the 1950s.[3] Those who regard the employment effects of microelectronics as largely neutral or beneficial are inclined to stress the historical analogy between this earlier period—indeed, all previous periods characterized by major technological breakthroughs—and the present spread and application of microelectronics technology, arguing that, as in the past, the "micro-revolution" will spur Western industrial economies to even higher levels of investment, overall employment, and growth.[4]

While most of these writers and researchers accept the fact that the introduction of new microelectronics technologies will create a certain amount of transitional, or short-term, economic dislocation—the displacement of certain categories of workers in some industries—they believe that the revolutionary character of these technologies will rapidly expand the demand for new products, create new markets, and increase the aggregate level of employment. They project new and growing employment opportunities in the production of semiconductors and related devices, in the manufacture of a very broad range of new products—video games, home computers, videorecorders, and digital watches—in programming, in repair services for and the installation of new products, and in various auxiliary activities such as the manufacture of furniture to support new electronic systems. The key to the success of the "micro-revolution" with regard to employment prospects is, clearly, that the new technologies so dramatically reduce costs that they transform luxury goods into mass market goods and create viable markets for entirely new goods.[5]

Those with opposing views in the debate regard historical analogies as misleading, arguing that the United States and other advanced industrial countries have, within the past two decades or so, entered an economic environment that is generally antithetical to high rates of employment growth, quite apart from the "micro-revolution."[6] A key term for those who find the employment effects of microelectronics alarming is "jobless growth,"[7] which means that productivity—and total production—continue to grow in an economic sector while employment levels remain the same or actually decline. Such "jobless growth" has occurred in the agricultural sector of advanced industrial economies for some decades. Such a condition has recently begun to develop in the manufacturing sector of these economies, particularly in the larger, more traditional industrial subsectors. Were it not for extremely high growth rates in employment in the various service sectors—government, education, health care, wholesale and retail trade, finance, insurance, and real estate—over the past two decades, there would have been much higher levels of unemployment in these advanced societies.[8] Most labor market analysts believe that the higher growth rates in these service sectors cannot continue. Employment growth in government and education has slowed significantly. Other service sectors continue to exhibit healthy employment growth rates, but these sectors are expected to expand more slowly in the future.[9]

If widespread "jobless growth" in the industrial countries is the product of the rapid introduction of advanced, labor-saving technology, then, according to those pessimistic about the employment effects of microelectronics, these countries are in for serious, long-term unemployment problems because of the pervasiveness of microelectronics and microelectronics-based technologies and the demonstrated substantial reductions in labor requirements attending the introduction of these technologies.[10] Widespread automation is expected to occur in the tertiary or service sectors of the economy in the advanced industrial societies—sectors that have provided the chief bases for employment expansion in the past two decades. Coupled with declining employment in agriculture and stagnant employment in the aggregate in manufacturing, the "micro-revolution" could be, as one author has noted, "the straw that will probably break the camel's back."[11]

There is much greater agreement on the employment prospects in the microelectronics industry itself over the next decade or so. Analysts who have closely examined the potential employment impact of the "micro-revolution" have reached essentially the same conclusion as that offered by the Organization for Economic Cooperation and Development in 1978: that "the electronics complex during the next quarter of a century will be the main pole around which the productive structures of the advanced industrial societies will be reorganized" and that countries that do not vigorously pursue the development and implementation of the new microelectronic technologies will suffer greater employment difficulties than those that do.[12]

Employment and Earnings in the U.S. Semiconductor Electronics Industry

Trends in Employment

Discussions about "the microelectronics industry" are frequently unspecific about the boundaries of that industry. In order to study trends in and characteristics of employment in microelectronics as an industry, it is essential to settle on a definition that simultaneously permits analysis and does not do violence to experts' sense of what the industry is and does. Some analysts consider the microelectronics industry to include all industries (and firms) that are routine producers and users of integrated circuits. At a minimum, this group of industries includes electronic components producers, manufacturers of computers, and manufacturers of scientific and measuring instruments. An informal survey compiled early in 1982 by the Division of Policy Development in the Department of Administration in North Carolina found that about 350 firms in the state either produced or used integrated circuit technology. These firms were spread across six industry sectors but were highly concentrated in two: nonelectrical machinery (Standard Industrial Classification 35), which includes the electronic computing equipment industry, and electrical and electronic machinery and equipment (SIC 36).[13] The difficulty with this characterization is that the microelectronics industry will continue to increase in size with the continuing incorporation of integrated circuits and microprocessors into the full range of U.S. industries. In this sense, for example, the American textile industry might eventually become a "segment" of the microelectronics industry.

The driving force of the "microelectronics revolution" is the design, production, and application of high-density integrated circuits. The industry category that captures the greatest share of this activity—and, in fact, is the most widely understood specification of the microelectronics industry—is the semiconductor electronics industry, classified as SIC 3674.[14] As noted by Sampson, Bourgeois, and Stein (chapter 2, this volume), SIC 3674 (semiconductors and related devices) is a subsector of 367, which is electronic components and accessories. There are eight other industry subsectors of SIC 367, including manufacturers of a variety of products ranging from radio and television receiving tubes to electronic connectors, transformers, resistors, and capacitors. In turn, SIC 367 is one of eight sectors of SIC 36, electrical machinery and electronic machinery and equipment. These industry sectors produce a wide range of products; some of the sectors (communications equipment, for example) are considered "high-tech" producers, but others (like household appliances) are not.[15]

As table 1 indicates, employment in the microelectronics industry—semiconductors and related devices—was about 226,700 in 1982: about 0.3 percent of all nonagricultural employment in the United States and just 1.2 percent of total manufacturing employment. Even if employment in the semiconductor electronics industry were to increase threefold over the next five to ten years, with only modest growth in employment elsewhere, the industry would constitute less than 1 percent of total nonagricultural employment in the United States. Moreover, nearly 90 percent of workers in the electronics industry are not engaged in the production (or supervision of production) of semiconductors. Though employment in electronic components and accessories (SIC 367) constituted well over a quarter of all employment in the electronics industry (SIC 36), the 226,700 workers in semiconductor electronics

Table 1. Employment in the Microelectronics Industry, 1982

	All Employees	Production Workers
Total nonagricultural	89,596,000	
Manufacturing	18,853,000	12,790,000
Electrical and Electronic Equipment (SIC 36)	2,015,500	1,216,600
Electronic Components and Accessories (SIC 367)	560,000	319,800
Semiconductors and Related Devices (SIC 3674)	226,700	90,000

Source: U.S. Department of Labor, Bureau of Labor Statistics, Supplement to Employment and Earnings, United States, 1909–1978 (Washington, D.C., July 1983).

(SIC 3674) represented only about 11 percent of that total. Even an exponential increase in employment in the semiconductor electronics industry would not by itself constitute an "employment revolution" or solve the problem of growing structural unemployment in the United States.

For statistical purposes the Bureau of Labor Statistics and various other public agencies, state and federal, have classified the microelectronics industry as one of a number of high-technology industries on the basis of certain characteristics.[16] For instance, the Bureau of Labor Statistics defines as high-technology industries all three-digit SIC industries that have at least twice the average proportion of total employment in scientific and technical occupations for all manufacturing *and* at least twice the average proportion of net sales devoted to research and development for all manufacturing. In the Bureau's view, industries with such characteristics are typically those "that generate a steady stream of new products as a result of technological advancement."[17] (The Bureau has also designated separate categories for technology-intensive manufacturing industries and high-technology service industries.)

As illustrated in tables 2 and 3, SIC 367 (electronic components and parts) is treated by the Bureau of Labor Statistics (and others) as a surrogate category for the microelectronics industry. This means that some of the evidence presented in the following sections of this chap-

Table 2. High-technology Industries

High-technology Manufacturing Industries[a]	
SIC 283	Drugs
SIC 357	Office, computing, and accounting machines
SIC 366	Communication equipment
SIC 367	Electronic components and accessories
SIC 372	Aircraft and parts
SIC 376	Guided missiles and space vehicles and parts
SIC 381	Engineering, laboratory, scientific, and research instruments
SIC 382	Measuring and controlling instruments
SIC 383	Optical instruments and lenses
Technology-intensive Manufacturing Industries[b]	
SIC 28	Chemicals and allied products (except SIC 283, drugs)
SIC 291	Petroleum refining
SIC 348	Ordnance and accessories
SIC 351	Engines and turbines
SIC 355	Special industry machinery, except metalworking machinery
SIC 361	Electrical transmission and distribution equipment
SIC 362	Electrical industrial apparatus
SIC 365	Radio and TV receiving equipment, except communication types
SIC 369	Miscellaneous electrical machinerh, equipment and supplies
SIC 384	Surgical, medical, and dental instruments and supplies
SIC 386	Photographic equipment and supplies
High-technology Service Industries[c]	
SIC 737	Computer and data processing services
SIC 7391	Research and development laboratories

[a]At least twice the average proportion of total employment in scientific and technical occupations for all manufacturing <u>and</u> at least twice the average proportion of net sales devoted to research and development for all manufacturing.

[b]Between one and two times the average proportion of workers in scientific and technical occupations for all manufacturing, with significant expenditures on research and development.

[c]Industries that do not produce high-technology products but that may significantly contribute to the development of these products through their research and development activities and computer services.

Source: U.S. Department of Labor, Bureau of Labor Statistics, Office of Economic Growth and Employment Projections, "'High-tech' Employment Growth (Draft)" (Washington, D.C., 1983).

Table 3. Characteristics of High-technology Industries

	Percentage of all workers employed as scientists, engineers, engineering and science technicians, and computer specialists[a]	R&D funds as % net sales (1980)[b]
Manufacturing, total	6.3	3.1
Drugs	18.4	6.2
Office, computing, and accounting machines	27.2	12.2
Communication equipment	24.8	8.6
Electronic components and accessories	14.9	7.6
Aircraft and parts	16.9	
Guided missiles and space vehicles	37.7	11.6
Engineering, laboratory, scientific, and research instruments	21.4	
Measuring and controlling instruments	16.1	6.1
Optical instruments and lenses	24.1	5.9

[a]Source: U.S. Department of Labor, Bureau of Labor Statistics, The National Industry-Occupation Employment Matrix, 1970, 1978, and Projected 1990, vol. 1, Bulletin 2086 (Washington, D.C., April 1981).

[b]Source: U.S. Department of Labor, Bureau of Labor Statistics, Office of Economic Growth and Employment Projections, "'High-tech' Employment Growth (Draft)" (Washington, D.C., 1983). These data are based on company reporting rather than establishment reporting that is used to classify the employment data and therefore are not entirely comparable. Data for optical instruments and lenses also include data for other instrument industries.

ter pertains to SIC 367, though data for the semiconductor electronics industry (SIC 3674) have been included whenever possible. Unfortunately some employment and occupational data are only available for the broader industrial classification of microelectronics. The problem is much more severe with data for North Carolina. Because the state has a very small number of semiconductor electronics firms and only a modest number of firms classified more generally within the electronic components industry, virtually all analysis for the state ranges between

SIC 367 and SIC 36, which contains industry sectors (and firms) that are not even classified as "high-technology." (Of the eight three-digit industry sectors under SIC 36, only two are classified by the Bureau of Labor Statistics as high-technology manufacturing industries; four are classified as technology-intensive manufacturing industries. See table 2.)

For the most part the present analysis focuses on employment and earnings in the merchant firm segment of the semiconductor electronics industry. Practically speaking, this is because industry data for employment and earnings are organized accordingly. (The most recent estimates suggest that merchant firms produce about three times as many integrated circuits as captive firms do.)[18] Table 4 presents the substance of the discussion below.

Between 1972 and 1982 total wage and salary employment in the United States increased by about 23 percent, from some 73,675,000 to 90,570,000 workers (table 4). During this ten-year period, aggregate manufacturing employment actually declined slightly. By contrast, growth rates among high-technology industries were remarkably high. Though their base was relatively small, high-technology manufacturing industries as a whole increased their employment by 41 percent, from 2,061,100 to 2,905,000. Though the growth rate of electronic components and accessories (SIC 367) was not the highest among the high-technology group, its rate of change, 60 percent, greatly exceeded the average for the group and was more than two and one-half times the growth rate of total wage and salary employment in the United States for the same period. Employment in the semiconductor electronics industry (SIC 3674) expanded from 115,200 to 226,700 (table 1), an increase of almost 100 percent. Thus employment in the semiconductor electronics industry grew at over four times the rate for total wage and salary employment and over twice the rate for employment in the high-technology manufacturing group. It is worth noting that within the general high-technology manufacturing category, the two industry sectors that experienced higher rates of growth—office, computing, and accounting machines, and measuring and controlling instruments—are often considered segments or extensions of the microelectronics industry (SIC 3674).[19] Because technological development in the semiconductor electronics industry is the catalyst behind new applications in computers, instruments, and communications technology, as well as in the semiconductor industry itself, it is considered "the engine of high-technology industry."[20]

Impressive as these growth rates are, they nevertheless did not transform the employment structure of United States industry or even of

Table 4. Employment Growth in High-technology Industries, 1972-1982 (in thousands)

Industry	1972	1982	% Change
Total Wage and Salary Workers	73,675.0	90,570.0	22.9
Manufacturing	19,151.0	19.031.0	-0.6
High-technology total	4,468.9	5,694.1	27.4
Group I: High-technology Manufacturing	2,061.1	2,905.0	40.9
Drugs	159.2	199.7	25.4
Office, computing, and accounting machines	259.6	490.0	88.8
Communication equipment	458.4	555.3	21.1
Electronic components and accessories	354.8	568.3	60.2
Aircraft and parts	494.9	611.8	23.6
Guided missiles and space vehicles	92.5	127.4	37.7
Engineering, laboratory, scientific, and research instruments	64.5	75.7	17.4
Measuring and controlling instruments	159.6	244.3	53.1
Optical instruments and lenses	17.6	32.5	84.7
Group II: Technology-intensive Manufacturing	2,191.3	2,269.6	3.6
Chemicals and allied products, except drugs	850.0	874.4	2.9
Petroleum refining	151.4	169.0	11.6
Ordnance and accessories	81.9	75.1	-9.1
Engines and turbines	114.6	114.9	0.3
Special industry machinery, except metalworking	176.9	179.4	1.4
Electric transmission and distribution equipment	128.4	110.2	-16.5
Electrical industrial apparatus	209.3	211.8	1.2
Radio and TV receiving equipment	139.5	94.5	-47.6
Miscellaneous electrical machinery	131.7	141.1	7.4
Surgical, medical, and dental instruments	90.5	160.5	77.3
Photographic equipment and supplies	117.1	138.4	18.2
Group III: High-technology Service	217.4	520.1	139.2
Computer and data processing services	106.7	357.4	235.0
Research and development services	110.7	162.7	47.0

Source: U.S. Department of Labor, Bureau of Labor Statistics, Office of Economic Growth and Employment Projections, "'High-tech' Employment Growth (Draft)" (Washington, D.C., 1983).

the manufacturing sector during the decade. As a proportion of all wage and salary employment, total high-technology manufacturing employment rose only very slightly, from 2.8 percent to 3.2 percent. Employment in the electronic components industry (SIC 367) rose from 0.5 percent to 0.6 percent of all industrial employment in the same interval, and from 1.9 percent to 3.0 percent of total manufacturing employment. If calculated by comparing the raw employment figures cited above with the broader industry figures that appear in table 4, employment in the semiconductor industry (SIC 3674) grew from a mere 0.2 percent to 0.3 percent of total nonagricultural employment over the decade. This apparent stability despite large growth rates can be attributed to (1) the relatively small total employment in high-technology manufacturing compared to overall industry and (2) the relatively rapid expansion of the major nonmanufacturing sectors of the United States economy (excepting agriculture, mining, and construction).

In contrast, employment in the services sector of the economy not only was approximately six times greater than total high-technology manufacturing employment in 1972, it grew much more rapidly over the decade. Service employment increased from 12,276,000 to 19,064,000 over the period, a change of 64 percent. Indeed, all major service-producing employment sectors—services, transportation and public utilities, wholesale and retail trade, finance, insurance, real estate, and government—were larger than the entire high-technology manufacturing sector in 1972 and grew more rapidly over the following decade.[21]

Employment Projections

There are no long-term employment projections for the semiconductor electronics industry (SIC 3674). However, there are a number of sets of industry projections for two- and three-digit SIC industries, including the electronic components industry. Table 5 gives projections for the set of high-technology industries defined in table 2. Depending on the trend or "scenario" chosen, total wage and salary employment is expected to grow from 92,611,200 in 1980 to between 110,053,200 and 117,425,200 in 1990, a range of growth from 19 percent to 27 percent. Manufacturing employment is projected to increase somewhere between 15 percent and 24 percent. In comparison, the Bureau of Labor Statistics projects the range of growth for high-technology industry employment as a whole to fall between 20 percent and 30 percent between 1980 and 1990, approximately the same rates projected for total wage and salary employment and somewhat above those for to-

Table 5. Projected Employment Growth in High-technology Industries, 1990 (in thousands)

Industry Group	1980	1982
Total wage and salary workers	92,611.2	90,570.0
Manufacturing	20,287.2	19,031.0
High-technology total	5,694.8	5,694.7
Group I: High-technology manufacturing	2,842.0	2,905.0
Group II: Technology-intensive manufacturing	2,385.4	2,269.6
Group III: High-technology service	467.4	520.1

Source: U.S. Department of Labor, Bureau of Labor Statistics, Office of Economic Growth and Employment Projections, "'High-tech' Employment Growth (Draft)" (Washington, D.C., 1983).

tal manufacturing employment. Projections specifically for high-technology manufacturing are likewise very similar to those for total wage and salary employment. The effects of the 1981–82 recession create a somewhat different set of projections for 1982–90. Because high-technology manufacturing industries were not affected as severely by the recession as total wage and salary employment and manufacturing as a whole, their post-recession growth rate is projected to be lower for this period.[22]

Three federal government economists have developed and evaluated employment projections through 1995 for a more broad-ranging group of high-technology industries. Group I, the most broadly defined, includes all three-digit SIC industries with a proportion of "technology-oriented" workers at least 1.5 times the average for all industries. Group II, the most narrowly defined, contains industries with a ratio of research and development expenditures to net sales at least twice the average for all industries. The intermediate group, Group III, includes manufacturing industries with a proportion of technology-oriented workers equal to or greater than the average for all manufactur-

1990 Low Trend	1990 High Trend I	Projected Change			
		1980–1990 Low Trend	1980–1990 High Trend	1982–1990 Low trend	1982–1990 High Trend
110,053.2	117,425.2	18.8	26.8	21.5	29.7
23,330.5	25,256.5	15.0	24.0	22.6	32.7
6,807.3	7,413.1	19.5	30.2	19.5	30.2
3,389.7	3,653.8	19.3	27.9	16.7	25.8
2,640.1	2,924.8	10.7	22.6	16.3	28.9
775.5	834.5	66.3	78.5	49.5	60.5

ing industries, and a ratio of research and development expenditures to sales at or above the average for all industries. It also contains two nonmanufacturing industries that provide technical support to high-technology manufacturing industries. For each of these three groups, high-technology employment is projected to grow more rapidly than total wage and salary employment under three alternative growth scenarios. This is true whether the base year chosen is 1980 or 1982. Despite these projected increases, though, high-technology industries, as defined under the three groups, are expected to increase their relative shares of total employment only modestly. For instance, under the most optimistic criteria projected employment expansion in Group I, the broadest classification of high-technology industry employment, would only increase the group's percentage of total employment from 13 percent in 1982 to 14 percent in 1995.[23]

In comparison, how is the microelectronics industry expected to perform? The Bureau of Labor Statistics has projected employment growth in the electronic components industry from 561,000 in 1982 to between 725,000 and 793,000 in 1990.[24] Under the most favorable assumptions, then, electronic components industry employment would increase by 42 percent; at worst, the industry would increase by about 29 percent. In general, then, the electronic components industry is projected to be a growth leader through the period even in comparison to high-technology industries as a whole.

It is possible as well to compare projected employment growth in the electronic components industry directly with projected increases in specific high-technology industries and non–high-technology industries for the period 1982–95. As table 6 indicates, electronic components is expected to be one of the fastest-growing industries in the United States between 1982 and 1995. The Bureau of Labor Statistics projects an average annual rate of change of 3.2 percent for the industry, which places it among the top ten most rapidly growing three-digit SIC industries for this period.

Over the period, this average annual rate of change would yield an aggregate increase of 52 percent. This compares quite favorably with the projected increase in the broadest category of high-technology employment (Group I) of 35 percent, and it is over twice the projected increase of 25 percent in total national employment during these years. Between 1982 and 1995 manufacturing employment as a whole is projected to increase about 22 percent.

On balance, then, the electronic components industry is expected to grow relatively rapidly over the next decade or so. Though a number of other high-technology industries are projected to grow even more rapidly—instruments and computers, for instance—the electronic components industry is expected to outperform the high-technology sector as a whole. Even at this relatively high rate of growth, the electronic components industry would account for only slightly more than 1 percent of the new jobs created between 1982 and 1995 under the "moderate" scenario proposed by the Bureau of Labor Statistics. Under this scenario, the national economy is projected to create about 25 million jobs between 1982 and 1995. Only some 289,000 of them are expected to be jobs in the electronic components industry. (Under the broadest definition, high-technology industries would account for 17 percent of all new jobs created between 1982 and 1995; under the narrowest, they would amount to slightly more than 3 percent.)[25]

Three-fourths of the new jobs are projected to be in service-producing industries (a very small number of which are high-technology industries), and fully one-third—34 percent—of the jobs under this scenario would be created in services industries proper. In other words, for every job created in the electronic components industry during these years, we should expect to see 65 new jobs in the service-producing

Table 6. Projected Employment Changes for Selected Industries, 1982-1995

Industry	Average Annual Rate of Change		
	1982-95	1982-90	1990-95
Fastest-growing			
Medical and dental instruments	4.3	3.2	6.1
Business services	3.9	4.1	3.6
Iron and ferroalloy ores mining	3.9	5.7	1.1
Computers and peripheral equipment	3.8	4.0	3.4
Radio and television broadcasting	3.8	4.2	3.0
Other medical services	3.8	3.6	4.0
Plastic products	3.5	4.1	2.4
Scientific and controlling instruments	3.4	3.2	3.7
Electronic components	3.2	3.6	2.7
New construction	3.1	3.3	2.8
Most Rapidly Declining			
Leather tanning and industrial leather	-3.3	-2.3	-4.9
Dairy products (processed)	-2.3	-2.1	-2.6
Wooden containers	-2.3	-2.3	-2.2
Leather products, including footwear	-2.2	-2.4	-1.9
Tobacco manufacturers	-2.1	-1.2	-3.4
Bakery products	-2.0	-1.0	-3.7
Railroad transportation	-1.6	-1.8	-1.2
Cotton	-1.5	-1.5	-1.5
Private households	-1.5	-1.9	-0.8
Dairy and poultry products (farm)	-1.3	-1.4	-1.3

Note: Data include wage and salary workers, the self-employed, and unpaid family workers.

Source: Valerie A. Personick, "The Job Outlook through 1995: Industry Output and Employment Projections," U.S. Department of Labor, Bureau of Labor Statistics, Monthly Labor Review, November 1983, 29.

sector of the economy (and 30 in the services industries alone). To carry the comparison one step further, the national economy is projected to generate over five times as many new jobs in eating and drinking places as in the electronic components industry.

Moreover, the clear trend in all high-technology manufacturing industries, particularly the semiconductor electronics industry and those industries tied closely to developments in microelectronics, is toward increasing automation of production. Not surprisingly, some of the earliest and most significant applications of microelectronics in an industrial setting—a production process—have occurred in the industry itself. The fabrication process is already highly automated and will become more so. Most of the tasks involved in microelectronics fabrication are quite routine; all industry analysts expect them to be subjected to an even greater degree of automation. Even the design phase of production, which involves considerable skill, has become more highly automated in recent years. The computer-aided design (CAD) of circuits, which is still in its infancy as a research and production tool, should greatly accelerate the process. In recent years Japanese semiconductor firms have substantially increased the level of automation of their assembly operations, and it is expected that U.S. semiconductor electronics companies will follow their lead.[26]

Increased automation in the semiconductor electronics industry is expected to produce much higher rates of growth in output than in employment. Rising productivity, then, will combine with an initial small employment base to create only a modest number of jobs in the industry over the next decade or so.[27]

The Occupational Structure of the Industry

Perhaps understandably, many people assume that the great bulk of occupations in high-technology industries are "high-tech" jobs. In fact, though there are some definitional problems and some minor differences in opinion on the issue, most experts agree that occupations that clearly meet the definition of high-technology jobs—occupations that directly involve their holders in either developing or applying new technologies—are engineers, life and physical scientists, mathematical specialists, engineering and science technicians, and computer specialists. Some managers, many of whom began their careers in clearly defined high-technology jobs, are also involved in the development and applications of advanced technologies, but we have few reliable statistics about them.[28]

Though high-technology industries have, by definition, higher concentrations of these "technology-oriented" occupations, in fact *most* high-technology occupations likewise exist in industries outside the high-technology sector. Furthermore, employment in high-technology occupations as a whole is projected to grow substantially more rapidly between 1980 and 1995 than either total occupational employment or total employment in high-technology industries.[29] This suggests that the recruitment of high-technology industries is not the only viable economic development strategy for upgrading the technical skills of the labor force and improving the employment prospects of a resident population. States and areas could as well pursue a manpower development strategy that would seek to provide the local labor force with specialized skills for technology-oriented employment in selected industries, both high-technology ones and others.

The microelectronics industry has at least three relatively distinct phases of the manufacturing process: research and development or product design, wafer fabrication, and assembly. Not only are the three phases characterized by relatively different occupational mixes, they also tend to occur in separate facilities and quite often in different geographical settings. Division of labor and actual physical separation of activities exist in virtually all manufacturing industries, but the pattern in the microelectronics industry is among the most pronounced.[30] Because of the organization of published data, it is difficult to examine the precise differences in the occupational structures of these three phases of manufacture in microelectronics. Product design clearly requires a higher concentration of very highly skilled technology-oriented workers; the wafer fabrication process is characterized by the broadest mixture of occupations and skills; the assembly process is the most labor-intensive phase of the three and requires the largest proportion of unskilled and semiskilled workers. But reliable data on employment structure are available only for relevant industry categories as a whole—in this case, for electronic components (SIC 367).

Table 7 compares the occupational structure of the electronic components industry with structures of selected other high-technology and non–high-technology industries. Among the latter are four industries that are presently major employers in North Carolina: textiles, apparel, furniture, and tobacco manufacturing. First, it is apparent that in 1980 both electrical machinery and electronic components had much larger proportions of workers employed in technology-oriented occupations than did manufacturing industries as a whole, by a ratio of 1.5–1.8 to 1. Manufacturers of electronic components employed about three times the percentage of engineering and science technicians as manufacturers as a whole.

Second, the electronic components industry is strikingly different in its concentration of technology-oriented employment from industries like steel and automobiles, which historically have been among the most important manufacturing sectors of the American economy. Indeed, the percentage of engineers employed by the electronic components industry exceeds the *total* percentage of workers in technology-

Table 7. A Comparison of Selected Industry Staffing Patterns, Shown as Percentage of Industry Employment

	WHITE COLLAR[a]		
	All	Technically oriented	Enginee
High-technology Industries			
Manufacturing	31.4	9.9	9.9
Electrical machinery (SIC 36)	38.3	17.4	7.0
Electronic components (SIC 367)	37.7	15.0	7.2
Office, Computing and Accounting Machines (SIC 357)	66.3	27.3	12.0
Computer and Data Processing Services (SIC 737)	96.0	26.0	1.7
Non-high-technology Industries			
Blast furnaces and basic steel products (SIC 331)	17.7	3.9	1.8
Motor vehicles (SIC 371)	20.2	5.9	3.4
Textile mill products (SIC 22)	17.4	3.4	0.5
Apparel and textile products (SIC 23)	18.2	2.4	0.2
Furniture and fixtures (SIC 25)	21.2	2.8	0.5
Tobacco manufacturing (SIC 21)	24.5	5.2	0.6

Source: Richard W. Richie, Daniel E. Hecker, and John U. Burgan, "High Technology Today and Tomorrow: A Small Slice of the Employment Pie," U.S. Department of Labor, Bureau of Labor Statitstics, Monthly Labor Review, November 1983, 52; idem, The National Industry-Occupation Employment Matrix, 1970, 1978, and Projected 1990, vol. 1, Bulletin 2086 (Washington, D.C., April 1981).

oriented occupations in either of the two industries. Blue-collar workers comprised from 75 percent to 80 percent of total employment in these industries; in the electronic components industry the figure was about 60 percent.

Third, though the electronic components industry ranks higher than the more traditional manufacturing industries in percentage of employees in high-technology occupations, some other high-technology

WHITE COLLAR[a]					
Life and Physical Scientists	Mathematical Specialists	Engineering and Science Technicians	Computer Specialists	BLUE COLLAR[b]	SERVICE
0.4	--[c]	2.2	0.7	60.8	1.9
0.2	--[c]	4.9	1.0	60.0	1.7
0.2	--[c]	6.4	1.2	61.0	1.3
0.1	--[c]	8.8	6.5	32.7	1.0
0.1	0.3	2.7	21.2	3.4	0.6
0.2	--[c]	1.5	0.4	80.2	2.1
0.1	0.2	1.7	0.5	76.8	3.0
0.2	--[c]	1.0	0.2	80.6	2.0
--[c]	--[c]	0.2	0.1	80.8	1.1
--[c]	0.0	0.6	0.2	77.2	1.6
0.6	0.0	1.1	0.4	72.4	2.9

[a]Includes professional and technical employees; managers, officials, proprietors; sales workers; and clerical workers.

[b]Includes crafts and kindred workers; operatives; and laborers.

[c]Less than 0.1%.

industries are far ahead of it: the office, computing, and accounting machine industry had nearly twice the percentage of individuals employed in such occupations in 1980. Furthermore, the computer industry had only about one-half the percentage of blue-collar workers as the electronic components industry.

Computer and data processing services, a high-technology service industry, had a labor force that not only was much more involved in technology-oriented tasks, but was in addition almost entirely composed of white-collar workers. This is not particularly surprising, but it does help illustrate the wide range of possibilities. In addition, this demonstrates, again, the relatively high dependence of the electronic components industry on engineers and engineering and science technicians and its relatively meager employment of computer specialists, at least in relation to its total labor force. In the latter sense, the electronic components industry is more like manufacturing industries as a whole—or even the "sunset" industries—than a number of other high-technology industries.

Fourth, among the four selected traditional industries that are prominent in North Carolina, technology-oriented employees exceeded 5 percent only in tobacco manufacturing. In comparison to the electronic components industry, these industries' use of engineers and engineering and science technicians in particular was minimal. Textiles, apparel, and furniture had among the lowest concentrations of these kinds of workers of any industry sectors at the two- *and* three-digit SIC level in the U.S. economy. The electronic components industry had nearly as large a percentage of employees in technology-oriented occupations as the textile and apparel industries had in total white-collar occupations, including managers, officials, proprietors, sales workers, and clerical workers.

Finally, it is worth emphasizing that although the electronic components industry had a relatively large percentage of workers employed in certain categories of technology-oriented occupations, the overwhelming majority of its employees—fully 85 percent—were not in "high-tech" jobs. Some 62 percent were in either blue-collar or service occupations. These figures are probably higher specifically for the semiconductor electronics industry (SIC 3674), and certainly they are higher for the segments of the microelectronics industry that are heavily involved in fabrication or bonding and assembly work.

Currently, in establishments chiefly engaged in fabrication or assembly processes, up to 60 percent of the employees are occupied in routine, highly repetitive tasks. Advanced manufacturing of semiconductors requires approximately one-third skilled engineering staff and technicians and two-thirds low-skilled production workers.[31] According to a 1971 Bureau of Labor Statistics Occupational Survey, less than 10 percent of the production, maintenance, construction, and power-plant labor force in the semiconductor electronics industry actually held skilled occupations. Of the remaining 90 percent, the overwhelming majority were semiskilled. Most of these workers "acquired" their skills through on-the-job training. Usually, this training was not extensive; workers were required to learn a few specific skills, which they then applied repetitiously.[32] States and communities that are interested in attracting microelectronics industries should consider whether the local labor force can accommodate such occupational patterns—always keeping in mind that many lower-skilled jobs may eventually be "lost" to automation.

Earnings

Microelectronics industry employees in technology-oriented occupations, particularly engineers, scientists, and computer specialists, are extremely well paid and in many cases hold equity in their firms. Managers and other high-level officials appear to earn salaries that are at least the equal of—and in many cases superior to—the average for manufacturing managers and officials as a whole. The highly educated and very skilled professional, administrative, and technical staff of microelectronics establishments hold jobs and receive salaries and related benefits that are the envy of much of the white-collar work force in manufacturing and, likely, in other industries as well.[33] However, their relative concentration in any given microelectronics establishment depends on which phase of manufacturing that establishment is involved in. Research and development facilities have high concentrations of highly paid, salaried workers; assembly operations have few. Data on earnings for production workers would thus tell us little about the former but much that was important about the latter.

The earning data for production workers are most representative of the advanced manufacturing phase of the microelectronics industry, the wafer fabrication process. This is because most semiconductor firms have located the majority of their assembly operations overseas—mostly in East Asia—to take advantage of low labor costs. In countries like Taiwan, South Korea, Hong Kong, Malaysia, Singapore, and the Philippines, wages for these highly labor-intensive manual assembly activities have been, and remain, a fraction of those paid in assembly operations in the United States. In some instances, foreign laborers in the microelectronics industry have received less than $100 per month for full-time employment.[34]

The most detailed information on wages and salaries in the microelectronics industry is the United States Department of Labor's 1979 *Industry Wage Survey*, which, though somewhat dated, nevertheless provides a good baseline against which to gauge changes, or the lack thereof, in the earnings profile of the industry. In 1977 production workers (not including technicians) in the semiconductor electronics industry earned $4.52 per hour on the average. In comparison, the average hourly rate for production workers in the southern states was $3.91, about 13 percent below the national average. Production workers in the electronics industry (SIC 36) received wages that were almost equal to the average hourly earnings for manufacturing workers as a whole, but workers in the semiconductor electronics industry drew wages that placed them second to last in average hourly earnings among workers in the electronics industry subsectors.[35]

Since 1977, average hourly earnings in the electronic components industry sector and the semiconductor electronics industry subsector have increased more rapidly than the earnings levels for the wider electronics industry and for manufacturing industries as a whole. In 1977, average hourly earnings in the electronic components industry and the semiconductor electronics industry were, respectively, 79 percent and 88 percent of the average for manufacturing. Over the next five years, these gaps decreased significantly, particularly for production workers in the semiconductor electronics industry, who by 1982 earned hourly wages that were over 96 percent of the average for all manufacturing industries.[36]

Table 8 presents data that allow us to compare current earnings for production workers in the semiconductor electronics industry with those for production workers in other industries. In general, the electronics industry is not a high-wage industry. In 1982, workers in the industry earned slightly less than the hourly average for all manufacturing workers; the $8.21 hourly wage placed the industry sixth among the ten durable goods industries and eleventh among the twenty major manufacturing industry groups.

Production workers in the electronic components industry received hourly wages that were substantially below the average for manufacturing (indeed, their earnings level was below the average for the total private nonagricultural labor force). They drew only 84 percent of the average hourly wage for all manufacturing. Workers in the semiconductor electronics industry earned $8.12 per hour compared to $7.17 for electronic components workers as a whole, but their average wage still fell below the average hourly earning for all manufacturing and for the electronics industry.

At the three-digit SIC level, workers in microelectronics and related activities fared very poorly in comparison with workers in other high-technology manufacturing industries. Average hourly earnings for electronic components production workers were the lowest—by a considerable margin—for the high-technology manufacturing group. Including technology-intensive manufacturing industries, workers in electronic components ranked next to last—twentieth in a group of twenty-one—in average hourly earnings. Earnings in other high-technology industries ranged as high as 85 percent above earnings for production workers in the electronic components industry. Given the broad character of these industry categories, these differences are quite large. Production workers in the semiconductor electronics industry (SIC 3674) received hourly wages that were about 13 percent above the average for electronic components workers, but this earnings level in 1982 was still below that of workers in most high-technology manufacturing industries and very far below the average wages received by workers in a large number of the industries in the high-technology and technology-intensive manufacturing groups.

There are a number of possible reasons why wages in the microelectronics industry are not higher. First, the production segment of the industry has a very small proportion of skilled workers, even in comparison to the blue-collar work force of a number of more traditional manufacturing industries. On this basis alone, one would not expect a high average wage level for the industry. Second, less than 20 percent of the labor force of the semiconductor electronics industry is covered by union contracts, and until recently a very large share of semiconductor production occurred in relatively small firms.[37] Virtually all of the unionized work forces are in "older," larger firms based in the Northeast. Many analysts believe that the "underdeveloped" character of labor organization in the western United States was a significant factor in the location of the microelectronics industry there during the 1950s and remains a continuing attraction of the Sunbelt as a whole for owners and corporate officials considering expansion or relocation of facilities and the establishment of entirely new operations.[38] Through 1981, there were no union-organized semiconductor establishments in Silicon Valley (Santa Clara County) and few, if any, such establishments in the southwestern and southern regions of the United States, where much recent growth in employment and output in the semiconductor electronics industry has occurred.[39]

The 1979 Department of Labor study cited earlier found that semiconductor electronics production workers in unionized plants had average hourly earnings that were 61 percent above those in nonunionized plants. As well, the study documented that workers in large firms (250 or more employees) received wages that were approximately 25 percent above those of workers in smaller firms.[40] A recent study of wage dispersion in United States manufacturing industries and mining found that the semiconductor electronics industry exhibited the greatest wage dispersion of any of the industries and that nearly two-thirds

Table 8. A Comparison of Earnings in the Semiconductor Electronics Industry with Earnings in Other Industries, 1982

	Average Weekly Earnings	Average Hourly Earnings	Average Weekly Hours
Total private nonagricultural establishments	266.92	7.67	34.8
Manufacturing	330.65	8.50	38.9
Durable Goods			
Lumber and wood products	283.48	7.46	38.0
Furniture and fixtures	234.73	6.31	37.2
Stone, clay, and glass	354.0	8.86	40.0
Primary metal industries	437.34	11.33	38.6
Fabricated metal products	344.18	8.78	39.2
Machinery, except electrical	368.81	9.29	39.7
Electric and electronic equipment	322.65	8.21	39.3
Transportation equipment	450.36	11.12	40.5
Instruments and related products	322.38	8.10	39.8
Miscellaneous manufacturing industries	247.56	6.43	38.5
Nondurable Goods			
Food and kindred products	310.87	7.89	39.4
Tobacco manufactures	369.68	9.78	37.8
Textile mill products	218.63	5.83	37.5
Apparel and other textile products	180.44	5.20	34.7
Paper and allied products	389.58	9.32	41.8
Printing and publishing	324.63	8.75	37.1
Chemicals and allied products	407.36	9.96	40.9
Petroleum and coal products	546.99	12.46	43.9
Rubber and miscellaneous plastics products	302.94	7.65	39.6
Leather and leather products	189.39	5.32	35.6

Source: U.S. Department of Labor, Bureau of Labor Statistics, Supplement to Employment and Earnings, United States, 1909-1978 (Washington, D.C., July 1983).

table 8, continued

	Average Weekly Earnings	Average Hourly Earnings	Average Weekly Hours
High-technology Manufacturing			
Drugs (SIC 283)	373.19	9.08	41.1
Office, computing, and accounting machines (SIC 357)	325.92	7.93	41.1
Communication equipment (SIC 366)	386.72	9.62	40.2
Electronic components and accessories (SIC 367)	282.50	7.17	39.4
• Semiconductors and related devices (SIC 3674)	321.55	8.12	39.6
Aircraft and parts (SIC 372)	461.55	11.23	41.1
Guided missiles and space vehicles and parts (SIC 376)	448.26	10.96	40.9
Engineering, laboratory, scientiric, and research instruments (SIC 381)	340.98	8.44	40.4
Measuring and controlling instruments (SIC 382)	317.99	8.03	39.6
Optical instruments and lenses (SIC 383)	346.32	8.53	40.6
Technology-intensive Manufacturing			
Chemicals and allied products, except drugs (SIC 28)	--	--	--
Petroleum refining (SIC 291)	589.19	13.30	44.3
Ordnance and accessories (SIC 348)	360.90	9.00	40.1
Engines and turbines (SIC 351)	445.35	11.39	39.1
Special industry machinery, except metalworking machinery (SIC 355)	354.42	8.95	39.6
Electrical transmission and distribution equipment (SIC 361)	312.31	8.07	38.7
Electrical industrial apparatus (SIC 362)	320.77	8.31	38.6
Radio and TV receiving equipment, except communication types (SIC 365)	299.53	7.70	38.9
Miscellaneous electrical machinery, equipment, and supplies (SIC 384)	352.04	8.89	39.6
Photographic equipment and supplies (SIC 386)	433.37	10.57	41.0

of the dispersion was due to wage variation between, as opposed to within, plants.[41] This could mean that workers in the same occupations but in different plants (in different regions) of the country had varying earning levels due to distinctive management strategies or local labor organization. It could also indicate varying functional specializations of different plants across the country.

A third reason for the relatively low earnings level of semiconductor electronics production workers is the characteristic composition of the industry's labor force. Women and minorities make up a disproportionate share of the semiconductor electronics work force. A 1979 survey conducted by the PHASE project in California found that women amounted to over 70 percent of the total production work force in Santa Clara County and held an even higher percentage of the lowest-skilled jobs, such as assembler and laborer. In contrast, women filled less than 24 percent of the craft positions and only about 19 percent of the technician positions, which require the greatest training and skill.[42]

The survey also indicated that 40 percent of women in assembly jobs were from racial or cultural minorities, mostly Hispanics and Asians. As a group, minorities held about 38 percent of the assembly jobs and 50 percent of the laborer positions, but only 20 percent of the technician occupations.[43] There are no comparable national totals or percentages. However, the Bureau of Labor Statistics' *Employment and Earnings* data series indicates that women have been and are presently substantially overrepresented in the semiconductor electronics industry work force relative to their shares of total employment in manufacturing industries as a whole and in most other high-technology industries.[44]

The preceding evidence is consistent with institutional, or structural-stratification, theories of wage determination, which do not ignore human capital considerations but argue that such differential "investments" rarely constitute a sufficient explanation of the observed variations in earnings across industries, occupational groups, and sexual estates. The social characteristics and institutional arrangements of the semiconductor electronics industry—a predominantly low-skilled production workforce, a sizable proportion of small firms, a largely nonunionized labor force, an overrepresentation of women and minorities in the workforce, and a high concentration of employment in the Sunbelt—have, in conjunction, produced a wage level that does not compare favorably with either manufacturing industries as a whole or selected high-technology industries. These differences are best understood as the outcome, or product, of a set of interactions that take account of the economic segmentation of industries, the varying organizational power of occupational groups, and the stratification of society into sexual estates.[45]

North Carolina and Semiconductor Electronics: Implications and Prospects

North Carolina is one of the few states in the Southeast that has attempted to articulate a coherent high-technology industrialization policy and then followed up that stated policy with concrete actions, including the commitment of substantial state funds.[46] Despite the presence of the Research Triangle Park and the recent initiatives of the Hunt administration, North Carolina is not, however, among the current national leaders in high-technology employment. The state has experienced considerable economic diversification over the past decade, but its industrial structure is still backward in comparison to the nation as a whole and, in fact, in comparison to other states in the Southeast. Though employment in the more traditional, local-resource-based industries like textiles, apparel, and furniture has fallen sharply as a percentage of total employment, in 1982 these industries still accounted for more than 20 percent of the state's total private nonagricultural employment (not counting government employment). Together, employment in textiles and apparel constituted about 40 percent of North Carolina's total employment in manufacturing (see table 9). Nationally, the figures for these employment categories were 3.2 percent and 10.2 percent, respectively. The concentration of "low-tech" industry employment in North Carolina is even more pronounced than these figures suggest, because the state has a larger manufacturing labor force, as a proportion of total employment, than any other state.

Estimates of high-technology employment vary widely, depending upon which classification scheme is adopted. If data for North Carolina were arranged according to the categories used by the Office of Economic Growth and Employment Projections in the Bureau of Labor Statistics (as in table 2), state employment in high-technology manufacturing industries would be about 48,500. Total high-technology industry employment, including employment in technology-intensive industries and high-technology service industries, would be about 119,400. Thus, high-technology industries in North Carolina would have employed a little more than half the number of workers employed in the textile industry in 1982.[47]

Table 9. Industry Employment and Wages in North Carolina, 1982

Industry Group	Insured Units
Total, all private industry	122,437
Agriculture, forestry, and fishing	1,985
Mining and quarrying	135
Construction	14,705
Manufacturing	10,451
20 Food and kindred products	664
21 Tobacco manufacturers	31
22 Textile mill products	1,358
23 Apparel and other finished goods	847
24 Lumber and wood products	1,983
25 Furniture and fixtures	704
26 Paper and allied products	192
27 Printing, publishing, and allied industries	1,104
28 Chemicals and allied products	305
29 Petroleum refining and related industries	20
30 Rubber products	306
31 Leather and leather products	54
32 Stone, clay, glass and concrete products	410
33 Primary metal industries	102
34 Fabricated metal products	572
35 Nonelectrical machinery	1,030
36 Electrical machinery, equipment, and supplies	258
37 Transportation equipment	204
38 Measuring and controlling instruments	108
39 Miscellaneous manufacturing industries	199
Transportation, communication, and utilities	4,240
Wholesale and retail trade	46,242
Wholesale	12,141
Retail	34,101
Finance, insurance, and real estate	8,811
Service industries	35,858
Federal total, all industries	1,631
State total, all industries	1,772
Local total, all industries	1,150

Source: North Carolina Employment Security Commission, Labor Market Information Division, North Carolina Insured Employment and Wage Payments, 1982 (Raleigh, 1983), xii-xxvi.

Average Monthly Employment	% Total Private Employment	Average Weekly Earnings
1,925,273		$263.60
17,914	0.9	183.48
4,427	0.2	362.40
106,859	5.6	263.53
785,034	40.8	282.24
44,043	2.3	276.43
25,202	1.3	455.35
225,077	11.7	240.08
84,148	4.4	180.15
31,411	1.6	225.94
79,319	4.1	227.34
22,070	1.1	419.00
21,345	1.1	275.44
38,672	2.0	401.48
669	--	317.70
25,637	1.3	342.61
6,096	0.3	199.16
16,125	0.8	312.00
8,832	0.5	345.46
23,498	1.2	320.37
50,976	2.6	387.18
52,062	2.7	347.11
15,615	0.8	313.49
9,499	0.5	346.96
4,738	0.2	235.93
111,472	5.8	389.06
491,554	25.5	216.01
127,808	6.6	348.10
363,746	18.9	169.59
94,908	4.9	297.06
313,105	16.3	240.00
48,667		396.62
104,113		297.60
234,499		238.64

Since no definition of high-technology industries has been widely accepted, it is often useful to look at the range of possibilities. Using the more broad-ranging definition of high-technology industries cited earlier, North Carolina high-technology industrial employment would range from 44,800 to 241,000. Even with the most inclusive definition, only about 12.5 percent of North Carolina's private employment could be characterized as high-technology industry employment in 1982.[48]

A number of studies have compared levels of high-technology employment across the states, though the definition of what constitutes high-technology industries varies somewhat from study to study. One recent report on high-technology industry employment in the *Monthly Labor Review* ranked the top ten states on the basis of numbers and proportions of workers in high-technology industries. North Carolina did not appear in any of these rankings; the only southern state that did was Florida, which had the tenth highest absolute level of high-technology industry employment in the United States, according to the authors' definition.[49] Two studies have included North Carolina. Researchers for the Massachusetts Department of Manpower Development ranked North Carolina twenty-first of twenty-five states in percentage of manufacturing employment in high-technology industries in 1980; at 10.6 percent, North Carolina was well below the national level (18.1 percent) and far below the two leading states, Arizona with 40.5 percent and Massachusetts with 34.8 percent.[50] A review of the prospects for high-technology industrialization in the southeastern states prepared by the Federal Reserve Bank of Atlanta generally concurred with the findings of the Massachusetts study. Among the eight states surveyed in 1982 North Carolina ranked sixth, at 9.8 percent, in percentage of high-technology manufacturing employment and was below the regional average of about 13 percent. Only Georgia and Mississippi had lower percentages, and Florida's percentage was nearly three times greater.[51]

Though North Carolina does not rank particularly high in terms of numbers or percentages of workers in high-technology industries as a group or sector, the state does have a sizable electronics industry. In 1982, that industry employed about 52,000 workers, which made it the fourth largest manufacturing industry in the state after textiles, apparel, and furniture. A recent research report from the North Carolina Department of Commerce noted that North Carolina ranked first among the eight South Atlantic states (including Florida) in production workers, value added by manufacturing, and value of industry shipments for the electrical and electronic equipment category (SIC 36), with about one-quarter of the total regional employment and dollars in

Table 10. Employment Structure of the Electronics Industry in North Carolina, 1981-1982

SIC	Industry	1982	1981
36	Electrical machinery, equipment, and supplies	52,062	54,363
361	Electric transmission and distribution equipment	5,477	4,662
362	Electrical industrial apparatus	8,122	8,492
363	Household appliances	5,564	5,881
364	Electric lighting and wiring equipment	7,229	7,207
366	Communication equipment	14,774	15,282
367	Electronic components and accessories	5,101	6,344
	Other electrical equipment and supplies	5,795	6,495

Source: North Carolina Employment Security Commission, Labor Market Information Division, North Carolina Insured Employment and Wage Payments, 1982 (Raleigh, 1983), xii-xxvi.

value added and value of shipments.[52] A relatively large, thriving electronics sector virtually assures that a state will have some measurable segment of its labor force in high-technology employment, because, as noted previously, most of the industry subsectors are classified either as high-technology or technology-intensive industries.

Nationally, the electronic components sector (SIC 367) employed about 28 percent of the electronics industry's total workforce in 1982 (see table 1); in North Carolina it accounted for less than 10 percent of the industry, some 5,100 workers (table 10). The semiconductor electronics industry (SIC 3674) proper is very small in the state: in 1982 there were only six such firms located in North Carolina, and each employed fewer than 100 workers. Through 1984, there were five semiconductor electronics firms, employing between 300 and 400 workers.[53]

These totals do not, of course, include employment in the manufacturing of semiconductors by the state's captive suppliers. Unfortunately, we do not have any reliable state-level figures on the composition of employment for these firms. In addition, these totals do not take account of firms in the state that manufacture semiconductors but are not classified as SIC 3674 because they engage in other industrial activities, which are predominant.[54]

In any event, the level of employment in semiconductor electronics does not compare favorably with the totals from states like California or Massachusetts. For example, in 1982 Massachusetts, which has a labor force about the size of North Carolina's, employed over five times the number of workers in electronic components (SIC 367) as North Carolina did (the data for SIC 3674 were not available). Two leading competitors with North Carolina for new and expanding semiconductor electronics facilities, Texas and Arizona, have semiconductor electronics labor forces that number in the thousands.[55]

Sectoral employment in the electronics industry is quite dispersed in North Carolina. The largest concentration of workers in the industry in 1982 was in communication equipment (SIC 366, by classification not in fact properly electronics, but generally regarded as such by analysts), which employed about 15,000. The electronic components sector (SIC 367) employed about one-third as many and ranked smallest of all related sectors of electrical manufacture in the state (table 10).

A recent Office of State Budget and Management forecast of nonagricultural employment in North Carolina projected an annual average growth rate in the electronics industry of only 0.7 percent for 1982–92, which is less than half of the projected overall growth rate of 2 percent. Under this forecast, electronics industry employment would grow from 53,200 in 1982 to 57,200 in 1992, a net increase of 4,000 workers.[56]

Under this scenario, analysts project employment in electronic components to expand from 6,345 in 1981 to 6,894 in 1990.[57] However, this forecast may not be particularly reliable. The historical time series for employment in the industry is relatively short; the recorded performance of the industry sector is probably not a good guide to its future performance; and the employment base for the industry is relatively small. One or two major "unexpected" locations of semiconductor electronic firms would generate employment levels in the state that would completely overwhelm the statistical forecast.

In all likelihood, employment in semiconductor electronics will grow much more rapidly than statistical forecasts have tended to suggest. Recent surveys of corporate officials in the electronics industry have placed North Carolina among the five most favorable state sites for location and/or expansion.[58] In addition, corporate heads of a number of semiconductor electronics firms have expressed specific interest in locating in North Carolina. According to an official in the North Carolina Department of Commerce, currently there are at least five semiconductor firms still considering the possibility of locating plants in the state.[59]

Hewlett-Packard, which is not classified as a semiconductor electronics firm but has a large captive operation, purchased an option on a land parcel in Wake County in 1980 and is apparently continuing to plan for construction of a facility. General Electric, which officially opened its Microelectronics Center in the Research Triangle Park in June 1982, has 100 employees but expects to expand to perhaps 1,000 over the next five to ten years. In addition, in May 1982 Texas Instruments announced that it had purchased options on two land parcels, one in Wake County and one near Asheville. Finally, Dynamit Nobel AG of West Germany, an explosives, plastics, and chemicals company, began construction of a silicon wafer manufacturing plant near the Research Triangle Park in August 1984. Within ten years company officials expect to employ 1,000 workers. One corporate official noted that it was Dynamit Nobels' "perception that North Carolina is rapidly emerging as the eastern U.S. center for high technology."[60]

Geographically, both employment and plants in the electronics industry are heavily concentrated in three major regions of the state: the Triangle area, the Triad area (Greensboro, Winston-Salem, High Point), and the Metrolina urban region (Charlotte, Gastonia). In 1982, these three areas, along with Buncombe County (Asheville), harbored nearly three-quarters of the electronics plants in the state and an even higher percentage of the total electronics employment; the Triangle and Triad alone contained almost half of the state's electronics plants (see figure 1). All of the establishments that employed more than 1,000 workers were in these areas (or in counties adjacent to them), and the great majority of electronics plants employing between 500 and 1,000 workers were similarly situated. Two-thirds of the state's counties had either no electronics establishments or only one.

Certain areas in the state, then, may experience a buildup in semiconductor electronics employment. In particular, the Research Triangle area is likely to attract a significant share of the expansion or relocation of domestic semiconductor electronics production *if* the ongoing restructuring of the microelectronics industry leads to increasingly widespread decentralization of production. However, even a major buildup within the Triangle area would not have a dramatic effect on the overall state economy. No industry experts presently forecast the creation of a new microelectronics complex with the size and dynamism of Silicon Valley.

The consensus, drawn both from academic research and interviews with industry officials, is that wafer fabrication, or "advanced manufacturing," plants are the most likely candidates for location in North

Figure 1. Electronics Firms Operating in North Carolina

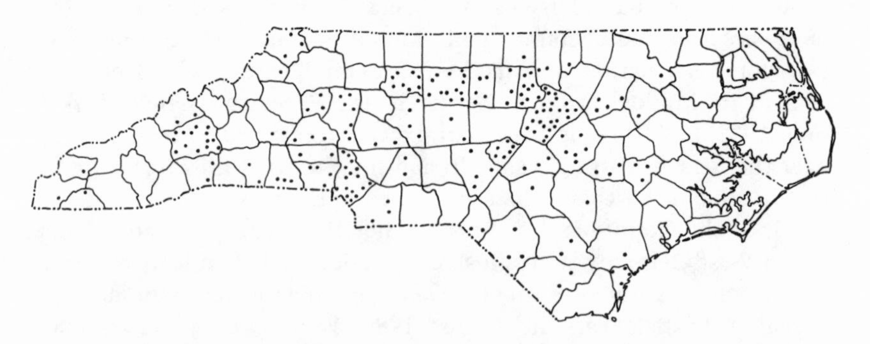

Source: North Carolina Department of Commerce, Economic Development Division, "The Electronics Industry in North Carolina," *Research Report*, May 1983, 5.

Carolina. Most corporate officials expect to keep their research and development facilities near headquarters, most of which are in California and Massachusetts.[61]

Assembly operations are less likely to move to North Carolina than wafer fabrication facilities because the low skill requirements of the former place a high premium on an abundant source of extremely cheap labor. According to most experts, neither North Carolina nor any other American state can compete with foreign locations, particularly Southeast Asian areas, on the price of labor. Recently, however, there have been some indications that the scale and complexity of the manufacturing and assembly processes associated with the development of very large-scale integrated (VLSI) circuits may lead to the location, and relocation, of more assembly operations in the United States.[62] The trend toward the increasing automation of production under VLSI circuit development has heightened labor force skill requirements and made necessary closer supervision and coordination of production by domestic actors. Depending on the strength of this trend, North Carolina could draw a notable segment of the "new" assembly operations as well.

Broadly speaking, neither the current projected growth of the semiconductor electronics industry in the state nor the more realistic scenarios of industry location and expansion would radically alter the state's current occupational structure or earnings structure and level. Though the semiconductor industry as a whole has a relatively high proportion of workers in technology-oriented occupations, particularly engineers and engineering and science technicians, the absolute increase in their numbers associated with the expansion of semiconductor electronics in any of its phases would not be great.

Because of the need for the advanced manufacturing phase of the semiconductor industry to be located near metropolitan areas, major universities, and large airports, it is likely that most future growth of the industry will occur in the Triad and Triangle areas. The location of even a relatively few new microelectronics plants in Raleigh–Durham, Greensboro–Winston-Salem–High Point, or even Charlotte–Gastonia would have a significant effect on the demand for labor. At present, almost any buildup of the industry would create some short-term imbalances in the labor market, particularly in the Triangle area, because of apparent existing insufficiencies in labor supply.

A 1980 survey of firms in the greater Raleigh area, including electronics firms, indicated that the greatest number—perhaps even half—of the anticipated skilled job openings between 1980 and 1985 would be in electronics or electronics-related positions. Local employers expressed great concern about the availability of adequately trained technicians. Two recent special reports on local job openings issued by the Labor Market Information Division of the North Carolina Employment Security Commission indicated that in both 1983 and 1984 the electronics technician and electronics mechanic/repairer positions were among the top ten most difficult positions to fill across much of the state. The chief reason was apparently a lack of adequately trained applicants.[63]

An interesting case is the recent experience of General Electric in North Carolina. Officials from General Electric announced in 1980 that the facility they planned to build in the Research Triangle Park would probably not employ a large number of local workers. In particular, they expected to hire few professionals and highly trained technicians from the state's labor force. Don Beilman, who was at the time vice-president and general manager for General Electric's Advanced Microelectronics operation and is presently President of the Microelectronics Center of North Carolina, stated at a 1981 university symposium that because North Carolina was not producing an adequate supply of engineers and electronics technicians, General Electric and other major electronics firms would continue to import a large share of their highly educated and skilled work forces from outside the state. In fact, General Electric hired North Carolina residents for only about one-third of the first hundred jobs associated with its new facility.[64]

The availability of skilled engineering, science, and other technicians is a more important consideration in decisions on where to locate high-technology facilities than the availability of scientists and professional engineers. In the case of scientists and engineers, high-technology corporate officials may bear with a local shortage if they are convinced that the local amenities—cultural, educational, and financial—will allow them to recruit needed professionals. The presumption is that the latter group is highly mobile, whereas the former is not.[65] Hence the concern among top-level officials in government and education that the current production of engineers in the state is short of the number needed would appear to be a less serious issue.

With sufficient funding, the state's community college system, one of the best and most comprehensive in the country, should be able to provide an adequate supply of skilled engineering and science technicians. Enrollments in the community college system's electronics engineering technology curriculum have been rising rapidly in recent years due in part to increased demand and, no doubt, in part to the publicity surrounding the establishment of the North Carolina Microelectronics Center. Currently most of the enrollment, and the sharpest increases in enrollment, have occurred in areas where semiconductor electronics plants are most likely to locate.[66] However, a number of major semiconductor plant locations would quickly strain the existing capacities of the community colleges in the areas in question. Even today, most have difficulty recruiting and retaining qualified instructors, and much of their training equipment has been made obsolete by recent technological changes in the electronics industry as a whole.

Training concerns aside, it is important to stress that the majority of employees in a typical semiconductor electronics work force will be either semiskilled or unskilled. North Carolina presently has an ample supply of these, particularly the unskilled, in the state. However, the largest concentrations of these workers, and unemployed workers as well, are generally outside areas where the semiconductor electronics industry is most likely to locate plants. A major buildup in the Triangle and/or Triad areas, then, would probably induce substantial immigration to these areas, leaving much of nonmetropolitan North Carolina as backward, or underdeveloped, in a relative sense, as before.

With regard to earnings, workers and managers highly skilled in the technology of the semiconductor industry receive relatively high wages and salaries, but their earnings are not characteristic of the great majority of workers in the industry. Production workers in the semiconductor electronics industry (or the electronic components industry) are not particularly well paid. Nationally, their earnings are very close to the average for manufacturing workers as a whole and are significantly below the average earnings of workers in the great majority of high-technology industry groups. The semiconductor industry is thus most accurately described as a "mid-wage" or "low-wage" industry.

The current evidence on microelectronics industry wages in North Carolina is both inapplicable and inconclusive because there are so few semiconductor production firms and workers in the state. Not only is there little wage information for SIC 3674 in North Carolina, the wage data for SIC 367 are not relevant because the composition of the electronic components industry in North Carolina is so different from its composition nationally. The confusion on the part of some state officials in North Carolina and others who have debated the relative wages of microelectronics workers is due in part to these considerations.[67] Any attempt to settle this dispute by comparing the wages of workers in other broad two-digit SIC groups suffers from the same problems. Presently, the best guides to the question of future wages in the microelectronics industry are national earning data and trends, such as those that have been presented here.

Notes

1. For instance, see Guenter Friedrichs and Adam Schaff, eds., *Microelectronics and Society: A Report to the Club of Rome* (New York: Pergamon Press, 1982); Colin Norman, *Microelectronics at Work: Productivity and Jobs in the World Economy*, Worldwatch Paper 39 (Washington, D.C.: Worldwatch Institute, October 1980); Tom Forester, ed., *The Microelectronics Revolution* (Cambridge: MIT Press, 1981).

2. Many of the best of these are European studies. See *The Impact of Microelectronics on Employment in Western Europe in the 1980s* (Brussels: European Trade Union Institute, 1979); see also Iann Barron and Ray Curnow, *The Future with Microelectronics* (London: Frances Pinter, 1979), and Clive Jenkins and Barne Sherman, *The Collapse of Work* (London: Eyre Methuen, 1979).

3. See, for instance, Eli Ginzberg, ed., *The Mechanization of Work* (New York: Scientific American, 1983).

4. One of the best summaries of this debate is still unpublished; see Paul Attewell, "Microelectronics and Employment: A Review of the Debate" (Santa Cruz: Department of Sociology, University of California, 1983). See also Alexander King, "Introduction: A New Industrial Revolution or Just Another Technology?" in Friedrichs and Schaff, *Microelectronics and Society*.

5. Norman, *Microelectronics at Work*, 7–15.

6. Attewell, "Microelectronics and Employment," 7–12.

7. Norman, *Microelectronics at Work*, 29–40.

8. Norman, *Microelectronics at Work*, 29–40; also King, "Introduction," 7–10.

9. King, "Introduction," 28–35; see also John Evans, "The Worker in the Workplace," 151–80, and Guenter Friedrichs, "Microelectronics and Macroeconomics," 181–202, both in Friedrichs and Schaff, *Microelectronics and Society*.

10. See Attewell, "Microelectronics and Employment," and Friedrichs, "Microelectronics and Macroeconomics."

11. Attewell, "Microelectronics and Employment," 7.

12. *Facing the Future: Mastering the Probable and Managing the Unpredictable* (Paris: Organization for Economic Cooperation and Development, 1979), 336.

13. Office of State Budget and Management, "Memorandum," table 5: North Carolina Employment by Sector and Selected Industries, January 1982.

14. The best discussion of the industrial structure of the microelectronics industry is Nico Hazewindus, *The U.S. Microelectronics Industry: Technical Change, Industry Growth, and Social Impact* (New York: Pergamon Press, 1982).

15. See "High Tech Employment Growth" (Draft, Bureau of Labor Statistics, Office of Economic Growth and Employment Projections, 1983).

16. "High Tech Employment Growth," 1–4.

17. "High Tech Employment Growth," 3.

18. Hazewindus, *The U.S. Microelectronics Industry*, 51–52.

19. In the United States, the computer industry is responsible for approximately 56 percent of the end use of semiconductors; see Giovanni Dosi, *Technical Change and Survival: Europe's Semiconductor Industry* (Sussex: University of Sussex, Sussex European Research Center, 1981), 23.

20. Donald L. Koch, William N. Cox, Delores W. Steinhauser, and Pamela V. Whigham, "High Technology: The Southeast Reaches Out for Growth Industry," *Economic Review* 68, no. 9 (September 1983): 6.

21. U.S. Department of Labor, Bureau of Labor Statistics, *Employment, Hours, and Earnings, United States, 1909–1984* (Washington, D.C., July 1984).

22. "High Tech Employment Growth," 8–10.

23. Richard W. Richie, Daniel E. Hecker, and John U. Burgan, "High Technology Today and Tomorrow: A Small Slice of the Employment Pie," U.S. Department of Labor, Bureau of Labor Statistics, *Monthly Labor Review*, November 1983, esp. 51–54.

24. Valerie A. Personick, "The Job Outlook through 1995: Industry Output and Employment Projections," U.S. Department of Labor, Bureau of Labor Statistics, *Monthly Labor Review*, November 1983, 30–32, table 5.

25. Personick, "The Job Outlook through 1995," 25, table 1; 30–32, table 5.

26. See National Academy of Engineering, *The Competitive Status of the U.S. Electronics Industry* (Washington, D.C., 1984), 5, 39–51; also see W. W. Goldsmith, "Bringing the Third World Home," *Working Papers* 9, no. 2 (1982): 24–31, and David O'Connor, "Changing Patterns of International Production in the Semiconductor Industry: The Role of Transnational Corporations" (Pre-

pared for the Conference on Microelectronics in Transition, University of California, Santa Cruz, 12–15 May 1983).

27. This is generally the case for high-technology manufacturing industries. See "America Rushes to High Tech for Growth," *Business Week*, 28 March 1983, 84–98.

28. See Richie, Hecker, and Burgan, "High Technology Today and Tomorrow," 54–55.

29. Eva C. Galambos, *Technician Manpower in the South: High Tech Industries or High Tech Occupations?* (Atlanta: Southern Regional Education Board, 1983), 1–2.

30. See Annalee Saxenian, "Silicon Chips and Spatial Structure: The Industrial Basis of Urbanization in Santa Clara County, California" (Working paper 345, Institute of Urban and Regional Development, University of California, Berkeley, 1981), esp. part III.

31. Saxenian, "Silicon Chips and Spatial Structure," 42–44.

32. Cited in Tom Bourgeois and Greg Sampson, "Sector Report: The Electronics Industry in North Carolina" (Working paper, Department of City and Regional Planning, University of North Carolina at Chapel Hill, 1981), part II.

33. See Saxenian, "Silicon Chips and Spatial Structure," 36–44; and Evans, "The Worker in the Workplace"; see also Craig Calhoun, "Computer Glamour and the Microelectronics Industry: Technological Change and Employment Prospects in the Continuing Industrial Revolution" (Working paper, Department of Sociology, University of North Carolina, 1984).

34. See Goldsmith, "Bringing the Third World Home," and O'Connor, "Changing Patterns of International Production in the Semiconductor Industry"; see also Linda Y. C. Lim, "Global Factors in the Global City: Labor and Location in the Evolution of the Electronics Industry in Singapore and Southeast Asia" (Working paper, Center for South and Southeast Asian Studies, University of Michigan, Ann Arbor, 1983).

35. U.S. Department of Labor, Bureau of Labor Statistics, *Industry Wage Survey: Semiconductors, September 1977*, Bulletin 2021 (Washington, D.C., 1979), 1–5.

36. U.S. Department of Labor, Bureau of Labor Statistics, *Supplement to Employment and Earnings, United States, 1909–1978* (Washington, D.C., July 1983).

37. See Bureau of Labor Statistics, *Industry Wage Survey: Semiconductors, September 1977*, 2.

38. Saxenian, "Silicon Chips and Spatial Structure," parts I and II; see also Andrew Goldenkranz and Joel Yudken, "Organizing the Unorganized Electronics," *Plowshare Press* 6, no. 2 (March–April 1981).

39. Saxenian, "Silicon Chips and Spatial Structure," part III; see also Goldenkranz and Yudken, "Organizing the Unorganized Electronics."

40. Bureau of Labor Statistics, *Industry Wage Survey: Semiconductors, September 1977*, 2.

41. See Carl B. Barsky and Martin E. Personick, "Measuring Wage Dispersion: Pay Ranges Reflect Industry Traits," *Monthly Labor Review* 104, no. 4 (1981): 35–41. Wage dispersion was measured by the coefficient of variation and the "index of dispersion"—the interquartile range divided by the median and then multiplied by 100.

42. Project on Health and Safety in Electronics, *Profile of the Electronics Industry Workforce in the Santa Clara Valley* (Mountain View, California, 1978).

43. Project on Health and Safety in Electronics, *Profile.*

44. See Bureau of Labor Statistics, *Supplement to Employment and Earnings, United States, 1909–1978.*

45. For a brief summary of the debate, and some evidence, see Robert Bibb and William H. Form, "The Effects of Industrial, Occupational, and Sex Stratification Wages in Blue-Collar Markets" (Working paper, Institute of Labor and Industrial Relations, University of Illinois at Urbana-Champaign, 1977); see also Randy Hodson, *Workers' Earnings and Corporate Economic Structure* (New York: Academic Press, 1983), esp. chap. 1–3.

46. For a description of the policies and associated funding, see Governor's Task Force on Science and Technology, report, vol. 2: *Economic Revitalization through Technological Innovation* (Raleigh: North Carolina Board of Science and Technology, 1984).

47. These data were collected by the Labor Market Information Division of the North Carolina Employment Security Commission under the ES–202 Program.

48. Total "technology-oriented" *occupational* employment would be higher, of course. This is because a large share—on average, over half—of all "high-tech" occupations are located in non–high-technology industries.

49. Richie, Hecker, and Burgan, "High Technology Today and Tomorrow," tables 6–8.

50. Massachusetts Division of Employment Security, Job Market Research, *High Technology Employment: Massachusetts and Selected States, 1975–1981* (Boston, July 1982).

51. Koch et al., "High Technology: The South Reaches Out for Growth Industry."

52. North Carolina Department of Commerce, Economic Development Division, "The Electronics Industry in North Carolina," *Research Report,* May 1983, 3.

53. North Carolina Employment Security Commission, Labor Market Information Division, ES–202 Program, 1982–84.

54. For a discussion of the basis of industry code assignment, see U.S. Department of Labor, Bureau of Labor Statistics, *Handbook on Standard Industrial Classification 1972* (Washington, D.C., September 1974).

55. U.S. Department of Labor, Bureau of Labor Statistics, *Supplement to Employment, Hours, and Earnings, States and Areas, Data for 1980–83,* Bulletin 1370–18 (Washington, D.C., August 1984).

56. North Carolina Office of State Budget and Management, Research and

Planning Services Section, *North Carolina Long-Term Economic-Demographic Projections* (Raleigh, February 1984), table 5.

57. These figures are based on unpublished detailed industry projections from the Labor Market Information Division of the North Carolina Employment Security Commission.

58. For instance, see the December 1983 survey of American electronics firms by Electronic Location File of Surrey, England, cited in *USA Today*, 21 January 1984, 1.

59. Interview with an anonymous official at the North Carolina Department of Commerce, 15 February 1985.

60. Thom Hill, "High-tech Plant to Provide 1,000 Jobs in Triangle Area," *Raleigh News and Observer*, 26 July 1984, C:1–2. See also Monte Basgall, "G.E. Dedicates Center of Microelectronics," *News and Observer*, 15 June 1982, A:16, and John Hinton, " 'Chips' Jobs Will Abound, Official Vows," *News and Observer*, 24 June 1982, A:21.

61. See Michael I. Luger, "Promises and Policies: The Economic Hope of the Microelectronics Industry," *N.C. Insight* 4, no. 3 (September 1981): 26–32.

62. National Academy of Engineering, *The Competitive Status of the U.S. Electronics Industry*, 1–7.

63. North Carolina Employment Security Division, Labor Market Information Division, *Special Report on the Analysis of Local Job Openings* (Raleigh, May 1983); see also *Special Report on the Analysis of Local Job Openings* (Raleigh, January 1985).

64. Ferrel Guillory, "New Infusion of State Funds Brightening Job Prospects," *New York Times*, 28 March 1982, section 12.

65. Robert Premus, *Location of High Technology Firms and Regional Economic Development*, Staff Study prepared for the use of the Subcommittee on Monetary and Fiscal Policy of the Joint Economic Committee of the U.S. Congress (Washington, D.C., 1982).

66. See North Carolina Department of Community Colleges, *Curriculum Registration Report by Curriculum* (Raleigh, 1980–81, 1981–82, 1982–83).

67. A good example of this misuse of wage information is Julie B. Hairston, "Computers Can't Save North Carolina," *The North Carolina Independent*, 24 June–7 July 1983, 1, 5–7.

11 Hazardous Wastes and Microelectronics in North Carolina

Carlisle Ford Runge

North Carolina faces the dual challenge of promoting economic development while maintaining a high level of environmental quality. Current state planning attaches high priority to the attraction of microelectronics firms as an industrial base for economic development. The adoption of such a policy raises several questions concerning the environmental impact of the technology as a generator of hazardous waste—some common to most industrial activities, others unique to high-technology and chemical-based industries.

Answering such questions with certainty is difficult, for three reasons. (1) Much of the information concerning potential hazards is highly technical. Those who understand the dangers may not try or be able to convey this information effectively to the public and to policymakers. Furthermore, many hazards involved in this industry are simply unknown to science. Technical and scientific problems involving long-term environmental hazards are therefore difficult to predict. (2) The microelectronics industry is relatively new. State governments have been quick to realize its potential economic benefits but slower in planning for and regulating possible environmental hazards. In North Carolina, as elsewhere, those most convinced of the value of drawing microelectronics firms to the state have little obvious incentive to raise issues concerning hazardous by-products of the new and promising technology. Even those who would regulate hazards may not be able to do so, because the technical information on which to base such regulatory policy is so scarce. The absence of clear government regulations creates further uncertainty inside and outside industry. (3) The greater the technical and scientific uncertainty, and the greater the problems of regulation, the higher the risks of unanticipated environmental damage.[1] These risks have been the primary focus of public concern.

If the scientific knowledge were available, effective regulations could potentially eliminate most environmental hazards. The key issue thus becomes better information and better communication of it. The search for better information can be organized around the scientific and regulatory issues listed above. First, what are the technical and scientific steps necessary to identify the hazards posed by the microelectronics industry? Second, what policies or regulations are most appropriate in dealing with them? Third, given the existing state of information, what are the risks associated with these hazards, and how do they compare with the net benefits of microelectronics development in North Carolina? Although complete answers to these questions will only emerge in time, it is important for government, industry, and the public to begin a process that will reduce the overall level of uncertainty about the environmental effects of North Carolina's microelectronics industry, so that economic development and the preservation of environmental quality may be seen as complementary, not competing, objectives.

Reducing Scientific Uncertainty

Although considerable attention has been devoted to the health hazards of working inside a microelectronics plant, less is known about environmental effects outside. State agencies in Silicon Valley and elsewhere in California have attempted to identify some of these effects. For example, beginning in 1976, the San Francisco Bay Area Air Quality Management District installed air filter strips near electronics firms between Santa Clara and Palo Alto to monitor five hazardous chemicals, including fluoride and chloride, with the object of determining whether high chemical concentrations in apricot trees in the district were an effect of residuals from microelectronics facilities. Relatively high concentrations of fluoride were in fact found in the center of the monitoring area, but the precise link from factory to air quality to trees was difficult to establish.[2] This inconclusive finding exemplifies a common, if difficult, problem of scientific uncertainty: incomplete information concerning the transmission mechanism from production processes in particular industries to overall levels of air, water, and ground pollution due to hazardous and toxic effluents.

Scientific uncertainty is not, however, simply a problem for scientists, but affects anyone involved in the management of hazardous materials. The Santa Clara fire marshal, for instance, has emphasized that "the electronics industry commonly uses hazardous chemicals that are flammable, explosive, poisonous and reactive when exposed to other chemicals." This presents "serious hazards" to firefighters or those in an area where a fire involving chemicals or a spill occurs. In addition he expressed concern about hazards arising from the transport of chemicals to and from microelectronics firms; as the number of these deliveries increases, so does the risk of accidents involving hazardous materials in transit.[3]

A major lesson from California's experience is that the technical and scientific information necessary to deal with these problems is often in short supply. The synergistic effects of interactions among the many chemicals used in wafer fabrication, as well as those hazardous and nonhazardous chemicals with which accidental spills or removal might bring them in contact, are not well understood. This is due in part to rapid advances in the industry's technology itself. As the Palo Alto fire marshal says, "They come up with new products faster than we can keep up with them."[4] The knowledge required to deal with interactions between new chemicals must grow at an ever-increasing rate.

Because it is further along the "learning curve," North Carolina can reap benefits from the experiences of those who have gone first. By studying the transmission mechanisms between industrial effluents and pollution, and the interactions of these effluents or pollutants with other chemicals, the environment, and humans, it should be possible to anticipate and plan for the sale and management of hazardous substances.

One area in which the need for information is imperative is public health. To this end, the Toxic Substance Project of the North Carolina Board of Science and Technology was established in 1980 at the direction of the governor to evaluate the most frequently used and most dangerously hazardous substances in industry and to develop comprehensive profiles on each. The project leader, Don Huisingh, stated quite simply that "there are questions pertaining to human health effects that we need to explore thoroughly. We want to be able to anticipate what problems there will be . . . before we have dead bodies."[5] Under Huisingh's direction, the project published complete management guides on fifty-one toxic substances in the spring of 1983.[6] These guides detailed known toxic chemical hazards, their sources, and their known interactions with other substances and agents. But many more such studies remain to be completed, and continued support for the project in an era of tight budgets is not assured, even though urgently needed.

Further specific attention should be focused on the microelectronics industry itself. One such effort is currently under way at the Research Triangle Institute (RTI) of North Carolina, the organization responsible for the compilation of the guides. These guides are part of a large project devoted to hazard assessment of the electronic component manufacturing industry. In addition to scientific and technical literature reviews and surveys of representative electronic component manufacturing facilities, much remains to be done, including analysis of newly introduced semiconductor materials such as gallium arsenide and indium antimonide. Moreover, RTI's work has focused primarily on in-plant hazards, not out-of-plant environmental impacts.[7]

A second area in which information is badly needed concerns the volume, location, and disposal of wastes. The volume of hazardous substances that will be generated in North Carolina by microelectronics firms depends on how quickly the industry expands and on the success of pollution prevention, recycling, and reuse strategies utilized inside and outside facilities. Some baseline estimates can be gleaned from data from California. Santa Clara County produced 1,849 tons of toxic wastes in a one-month period in 1980; 80 to 90 percent came from some of the five hundred electronics plants in the county.[8] Although that volume would not be achieved in North Carolina unless heavy development occurs, even a fraction of that amount requires careful monitoring and thorough information on the type, disposal method, and location of the wastes involved. As discussed below, North Carolina's institutional capacity to monitor these wastes, at any level, is open to question. The volumes of such hazards may be small, but volume alone is not an adequate indicator of harm: "The electronics industry wastes are not so great by volume, but they are some of the more nasty types."[9]

When the North Carolina General Assembly approved $24.4 million for the Microelectronics Center of North Carolina (MCNC) in the spring of 1981, a number of groups proposed the establishment of research capability to explore health and environmental hazards, as well as an advisory committee on environmental and occupational health. This committee was to monitor and assist the industry in the identification of disposal technologies and methods for reducing the total volume of waste. It was not funded, however, as part of the final bill. The absence of such capability means that less technical and scientific information can be related systematically to the problems and needs of industry.

To date, responsibility for providing much of this information has fallen to the universities and to industry itself. Several university and industry scientists have suggested areas in which scientific and technical information could be improved.[10] At the Interim Semiconductor Research and Fabrication Laboratory for MCNC many of the problems likely to be faced by industry in the disposal and recycling of hazardous materials are currently being met on a manageably small scale.

A scientist from the laboratory has identified the two main sources of difficulty as (1) acid neutralization and (2) solvent collection and disposal.[11]

Acid neutralization can occur both on premises and in publicly owned treatment works. The recently established General Electric Microelectronics Center can monitor levels of acidity in-plant and can neutralize up to two thousand gallons of sulfuric and hydrochloric acid per month at its Research Triangle facility in Durham County. What cannot be neutralized in-plant must be sent farther away to larger treatment facilities. General Electric's environmental health and safety specialist has remarked that area treatment facilities could become inadequate if microelectronics firms, with their large water volume requirements, continue to locate in the Research Triangle. Durham County regulations were "basically set up to control chicken feathers, fat and rags," not the hydrofluoric acid, metals, and solvent waste streams generated by the microelectronics industry.[12]

Hydrofluoric acid poses special problems. North Carolina law prohibits the imposition of control regulations that are more stringent then federal law.[13] Although fluoride is more toxic than chloride and other industrial toxic substances, the buffered hydrofluoric acid stream is not a federally designated hazardous waste. As a result, there is a "window" in North Carolina's regulatory policy that allows firms to dump this acid down the drain.[14] The federal Pretreatment Standard for the electronics industry, recently promulgated by the Environmental Protection Agency (EPA), also does not control fluoride concentration of effluents discharged to publicly owned treatment works.[15] Even so, at a cost of more than $600 per month General Electric voluntarily ships its hydrofluoric acid to Seaboard Chemical of Jamestown, Virginia, for precipitation treatment.

Where wastes such as fluoride cannot be controlled in-plant, reclaimed, or incinerated, a landfill must be developed that is capable of handling them. There has been controversy, however, over whether any landfill can contain such wastes over time. William Sanjour of EPA's Hazardous Waste Implementation Branch noted in late November 1982 that most, if not all, existing land disposal facilities for hazardous wastes are leaking contaminants and will later become imminent hazards to public health and the environment. Citing studies showing that leachate collection systems do not work and that all landfill liners eventually leak, Sanjour recommended that EPA ban the land disposal of all or most hazardous wastes.[16]

A ban on land disposal would place a premium on technical and scientific advances in pollution prevention and the reclamation, recycling, and reuse of hazardous materials. At the General Electric plant, for example, nitric acid contamination in the fluoride acid stream has created problems in the resale of hydrofluoric wastes.[17] Techniques to segregate these streams can improve the ability of industry to reuse or sell them when they are no longer needed. The development of such techniques can be accelerated if the search is assisted by state agencies, rather than left solely to industry.

Similar assistance can aid in the elimination of solvents, the second major area of concern. At the General Electric facility, many of the solvents used can be incinerated for BTU value; the company thus avoids using chlorinated or halogenated solvents, because these generate hydrochloric acid when burned. The first suspected contamination of groundwater from the microelectronics industry in North Carolina originated in an IBM semiconductor facility in the Research Triangle Park; chlorinated solvents were the major culprit.[18] Instead of chlorinated solvents General Electric uses xylene, isopropyl alcohol, acetone, and other nonchlorinated hydrocarbon solvents that can be readily recycled or safely burned in appropriate incinerator facilities.[19] If state agencies were prepared to assist industry with incineration technologies, or to promote nonchlorinated hydrocarbon rather than the use of chlorinated solvents, cooperation might significantly reduce risks of future environmental pollution.

In summary, improved technical and scientific information can assist in the identification and prediction of environmental hazards. Initial efforts by the Toxic Substances Project of the Research Triangle Institute to identify healthy and interactive effects of various substances should be continued and expanded, focusing directly on the microelectronics industry. Special efforts must identify how hazards are transmitted from factory to air, water, and ground. Information on the potential volume, location, and disposal of wastes should also be developed in cooperation with industry to monitor and control future treatment, reuse, or resale. Special effort should also be made to promote the prevention of waste generation through appropriate choices of acids and solvents. Current evidence suggests that landfills should be *last* in the line of options for disposal, and major attention should be devoted to in-plant recycling, reuse, and disposal.

Reducing Regulatory Uncertainty

State agencies are directly responsible for maintaining the state and local regulatory climate within which the microelectronics industry must operate. A common misperception of government officials responsible for attracting investment is that a "good regulatory climate" means little regulation of environmental hazards. From the point of view of industry, ambiguous or unenforced state and local regulations can lead to higher levels of uncertainty than strict but straightforward policies. Industry executives often find regulatory uncertainty far riskier in terms of implied future liability than less lax but more clearly defined public policies.

Three state laws control nonradioactive hazardous waste generation, discharge, transport, and treatment in North Carolina. The Solid Waste Management Act (N.C. Gen. Stat. § 130–166, Art. 13b) was amended in 1979 to place the Department of Human Resources (DHR) in charge of implementing all state and federal legislation on hazardous waste management and to give the Commission for Health Services responsibility for rules on solid waste management facilities. Under the act, DHR may delegate authority to municipalities or counties to perform any part of the state program within local jurisdiction. (Few if any such delegations have yet occurred.) Second, the Oil Pollution and Hazardous Substances Control Act of 1978 (N.C. Gen. Stat. § 143–215) assigns state authority for hazardous substance emergencies to the Department of Natural Resources and Community Development. Third, the Toxic Substances Act (N.C. Gen. Stat. § 14–284.2) makes it a felony to dump, incinerate, or dispose of certain specified substances except pursuant to this or other federal or state law. Overall, this patchwork of regulation has been judged seriously defective by at least one analyst, who stated that "the approach is piecemeal, lacking in long-term perspective, and still incomplete."[20]

The confusing array of agencies involved in this process led Governor James B. Hunt to appoint, in July 1980, his Task Force on Waste Management and ultimately to establish the Governor's Waste Management Board to oversee the many agencies involved. Efforts to implement this board's role as monitor have so far proved difficult, as evidenced by the continuing self-protective initiatives of firms in high-technology industries.

Several examples illustrate the difficulties here. In a seemingly innocuous episode at the General Electric plant, large quantities of sodium chloride (salt) were slated for disposal. Salt is not ordinarily considered "toxic," but increased salinity in water can have serious effects if the material is put directly into sewer drains. Environmental safety specialists at General Electric checked with local authorities, who in turn referred the industry to state-level agencies, who requested that the solution be trickle discharged.[21] Without General Electric's in-house decision to inquire, environmental costs might have resulted. Though minor in this case, such costs could have been much higher if more hazardous materials had been involved.

In a more serious case involving highly flammable photoresist materials, General Electric chose to incinerate rather than solidify the wastes for landfill disposal, despite the fact that the cost of incineration is from five to seven times greater than solidification.[22] The reason for the choice was minimization of future liability. However, General Electric's Research Triangle facility produces only three to four drums of photoresists per year. If it and other plants produced more, could they still afford to incinerate? Industry executives are not sure.

For this reason, companies involved in microelectronics are especially concerned with state policy toward hazardous waste landfills. This issue goes beyond the primary question of whether landfills can be made ready to receive wastes. Because of the dangers and potential liability associated with transport of hazardous wastes to landfills, and the virtual certainty that landfills will eventually leak—likewise leading to liability and damages—many experts believe both state and federal policy should aim to promote incentives for above-ground retrievable and detoxifiable storage, pollution prevention, recycling, and reuse. In recent hearings before the U.S. Congress, Clay Robinson, speaking for the National Hazardous Waste Treatment Council, noted that federal policy is promoting and subsidizing landfills that are inherently unsafe—which has had a "chilling effect upon the development of higher technology alternatives," such as incineration, stabilization, detoxification, recycling, and disposal.[23]

Rather than following the federal government's somewhat dubious lead, state government should provide alternatives that are actually more attractive to industry. This is consistent with a general shift in responsibility for hazards to the states.[24] As William Sanjour of EPA's Hazardous Wastes Implementation Branch puts it: "U.S. industry has the knowledge and technology to deal with the problem, but as long as cheap landfills are available and the government subsidizes it," alternative technologies will not be used. According to Sanjour, EPA's promises to provide emergency relief and other assistance when a landfill fails are analogous "to the Federal Aviation Administration issuing an airworthiness certificate to an airplane it knows to be unsafe on the basis that they promise to provide additional fire trucks and rescue vehicles and more hospital beds."[25]

General Electric's environmental health and safety specialist has recommended that both federal and state policy should aim to provide incentives for recycling and reuse technologies.[26] In North Carolina, these incentives could include not only tax breaks for the installation of disposal, recycling, and reuse mechanisms in high-technology facilities but also stronger local ordinances and regulations as well as economic incentives to reduce pollution at its source. A spokesman for Integrated Circuit Corporation has stressed that microelectronics companies' location decisions not only involve the often-quoted virtues of a good labor pool and quality of life but also depend on zoning regulations and local services that allow companies to operate with the assurance of a structure of local support.[27] So far such assurance does not exist in North Carolina. John T. Caldwell, former chancellor of North Carolina State University, remarked during debate over locating MCNC that there are "doubts as to whether we, at the present time, have a local government infrastructure and a citizen commitment of sufficient strength to plan for major growth" in high-technology development.[28]

Another example from California illustrates what Caldwell may have had in mind. In January 1982 it was disclosed that an underground disposal tank at the Fairchild Semiconductor facility in San Jose was leaking into the local water supply. Faced with more than $5 million in lawsuits thus far, the company had spent $12 million cleaning up the leak as of February 1983. This situation led to the passage in 1983 of a model waste disposal ordinance that provides a prototype for local action. The ordinance resulted from joint planning by a task force composed of local industry representatives, labor groups, and community organizations. Its key provisions required new double-walled chemical storage containers, regular monitoring of facilities for leaks, and companies' filing of hazardous material "impact statements" for public and fire department use.[29]

At the state rather than local level, a number of proposals for incentives favoring recycling and reuse have been proposed for North Carolina. One is the development of a "negative pollution tax" in which taxes would be levied on firms discharging hazardous substances for disposal, whereas firms that reduced hazardous discharges through recycling and reuse would be eligible for both tax reductions and subsidies.[30] "Negative taxes" create major incentives toward in-plant process modification to prevent wastes from being produced and entering the environment. A number of states have already established such schemes. In Wisconsin, machinery and equipment used for treating hazardous wastes is exempt from state property taxes. Oregon, which has the most extensive set of incentives, grants a 100 percent tax credit to industries for hazardous waste control facilities that produce either a usable source of energy or something of economic value from waste products; the credit can be taken either as a 10 percent write-off for ten years against income or excise taxes, or as a 5 percent credit per year for twenty years against ad valorem or property taxes.[31]

In North Carolina such schemes might well have promoted the use of hydrocarbon solvents reusable for BTU value and thus prevented the groundwater contamination episode at the IBM plant, in which chlorinated hydrocarbon solvents were the culprit. Indeed, linking tax incentives and subsidies to pollution prevention technologies provides a major opportunity to marry hazardous waste reduction to industrial recruitment. This was a major conclusion of a recent state conference organized around the theme "Pollution Prevention Pays."[32] One of the contributors, a professor of chemical engineering at North Carolina State University, remarked that "in most states it is clear that substantially more source control of hazardous waste can be implemented. The current challenge in the field of waste elimination is primarily the implementation of existing technology. At this stage with relatively little waste elimination, modest investments can typically yield very attractive savings from reduction or recovery of waste constituents."[33] The critical element from industry's point of view is the comparative cost of in-plant versus out-of-plant disposal. By reducing the cost of in-plant recycling and reuse through tax incentives and raising the cost of out-of-plant disposal through charges, proposals such as the "negative pollution tax" can foster a regulatory climate in which firms significantly reduce the volume of hazardous wastes that must be managed.

The rudiments of such a tax program are already in place in North Carolina. Real estate and certain equipment for waste disposal and resource recovery are already excluded from property tax, and parallel provisions allow for franchise tax exclusions. State law (N.C. Gen. Stat. § 130–166.18) also authorizes the Department of Human Services to develop standards for special tax classes for "recycling, reduction, or resource recovering facilities . . . or equipment." However, these provisions can be significantly expanded and linked directly to the needs of microelectronics and other high-technology industries.

In addition to tax incentives, direct efforts to advance technologies for in-plant reuse and recovery, including efforts at MCNC, can be continued and expanded. Where possible, supervision and monitoring by advisory committees, as proposed during initial legislative consideration of MCNC, as well as citizen and industry education, can help to reduce concerns and assist in providing additional expertise. State agencies such as the Department of Human Resources and Waste Management Board must pay greater attention to the zoning requirements, fire services, and other infrastructure necessary to maintain public safety. Proposals have recently been made for a state "pollution prevention center" with responsibilities for education, research, and development of both technical and policy alternatives.

In summary, both state and local policies can significantly enhance North Carolina's attraction for the microelectronics industry by responding to the industrial need for in-plant solutions to hazardous waste management. Tax and other incentives to adopt safe disposal methods can reduce the amount of wastes that ultimately must be transported and stored. Above-ground retrievable storage can then assume its proper role, with below-ground disposal as a last and least attractive option. Finally, strong local support services and monitoring, regulatory enforcement, and emergency response capabilities can increase both citizen and industry assurance.

Unfortunately, as the major state agency responsible for coordinating hazardous waste issues, the Governor's Waste Management Board has failed to seize the initiative required for such regulatory reforms. Because industrial recruitment has been perceived as competitive with hazardous waste management, rather than its complement, North Carolina continues to lag in the development of state and local regulations appropriate to advances in technological development, concentrating too much of its attention instead on landfill siting.

Conclusion

Improved scientific information, education, and regulatory reform can limit the overall risks of microelectronics facilities' wastes to human health and environmental quality in North Carolina. Without information on likely hazards and their effects, and regulations designed to prevent hazardous situations from occurring and local institutions that can cope with them if they do, North Carolina reduces its capacity to respond to an unanticipated episode of environmental damage. Whether such damage will occur is unknown, although the episode at IBM suggests that some risk exists, and the experiences of industry in California and elsewhere provide evidence that accidents can occur despite conscientious planning. Improved scientific information on the nature of potential hazards can assist not only in preventing their generation, but in dealing with spills or exposure if and when they occur. In addition, scientific information and general public education can improve the capacity of state and local regulation to define and assign liability. Incentives for firms to take care of wastes in-plant also minimize chances of wide public exposure to hazards. Finally, the development of local ordinances and services by well-trained personnel familiar with hazardous substance management can provide the institutional foundation for the development of safe, well-serviced facilities.

Scientific information on hazards in industry can be useful in a variety of contexts. It can assist industry in the development of alternative preventative and cost-saving recycling and reuse technologies. It can reduce health costs associated with industrial accidents in areas other than microelectronics. It can also inspire public confidence that through careful monitoring, the new- and high-technology industries, including microelectronics, will be a comparatively safe economic option. Regulations designed to promote in-plant reuse and recycling have obvious benefits to the public and industry alike, particularly because they often result in significant cost savings in production.[34] By reducing the net volume of waste, they also put less pressure on existing out-of-plant treatment and landfill facilities. Public investment in local services, such as fire services, monitoring, and emergency medical and spill clean-up capacity, as well as more advanced treatment facilities, can also create significant positive response in both the public and industry.

Will such information and regulations also discourage recruitment of large new microelectronics firms? The answer is unclear, but the views expressed by authorities from General Electric's Research Triangle plant lead one to doubt so. It is more likely that in the early stages of microelectronics recruitment, even a single episode injurious to human health or environmental quality can raise public alarm and allow rhetoric to overtake realistic assessment of hazards. If this occurs, many long-term benefits of industrial growth may be forgone. Regardless of the exact nature of risks and benefits, North Carolina can improve both the future welfare of its citizens and the attractiveness of its investment climate by reducing scientific and regulatory uncertainty in its environmental policies for hazardous waste management.

Notes

1. This problem is treated more formally in Carlisle Ford Runge, "Risk Assessment and Environmental Benefits Analysis," *Natural Resources Journal* 23 (July 1983): 683–96. See also Robert Wilson, "Risk Measurement of Public Projects," in *Discounting for Time and Risk in Energy Policy*, ed. Robert C. Lind et al. (Baltimore: Johns Hopkins Press, 1982), 205–49.

2. S. Yoachum and M. Malone, "The Chemical Handlers," special news reprint, *San Jose Mercury News*, 6–8 April 1980.

3. Yoachum and Malone, "The Chemical Handlers."

4. Yoachum and Malone, "The Chemical Handlers."

5. Quoted in Joseph T. Hughes, Jr., "Microelectronics: A Healthy Future for North Carolina," North Carolina Center for Public Policy, *North Carolina Insight* 4, no. 3 (September 1981): 33–38.

6. See *The North Carolina Toxic Substances Management Guide*, ed. and comp. Don Huisingh et al. (Raleigh: Toxic Substances Project, Office of the Governor, 1983), published as a loose-leaf binder containing the individual substance guides.

7. See Research Triangle Institute, *Hazard Assessment of the Electronic Component Manufacturing Industry* (Research Triangle Park, March 1981).

8. Yoachum and Malone, "The Chemical Handlers."

9. David Storm, regional head of the California agency responsible for managing hazardous wastes, quoted in Hughes, "Microelectronics," 38.

10. John Hauser, head of the Interim Semiconductor Research and Fabrication Laboratory of MCNC, headquartered in the Electrical Engineering Department of North Carolina State University in Raleigh, and Carol Scott, the environmental health and safety specialist, General Electric Microelectronics Center, Research Triangle Park, N.C., were both interviewed by telephone during preparation of this chapter. Don Huisingh, also of North Carolina State University, provided many valuable materials and insights.

11. John Hauser (see note 10).

12. This is readily confirmed by inspection of the Sewer Use Ordinance, City of Durham, North Carolina. See especially Article II, "Permissive and Prohibited Use of Public Sewers," Division 1, Sections 18–73.

13. This results from the so-called Hardison Amendment, a measure that prevents North Carolina from exceeding the standards developed under federal statute.

14. Carol Scott (see note 10).

15. See *Code of Federal Regulations*, Title 40, Part 469.

16. See "Panel Told All Landfills Leak, EPA Rules on Hazardous Waste Land Disposal Inadequate," *Environment Reporter*, 3 December 1982, 1277.

17. Carol Scott (see note 10).

18. Carol Scott (see note 10).

19. Carol Scott (see note 10).

20. Terrence Pierson, "A Strategy for Siting Hazardous Waste Management Facilities: The North Carolina Case" (Working paper, Department of City and Regional Planning, University of North Carolina, Chapel Hill, 1981), 10.

21. Carol Scott (see note 10).

22. Carol Scott (see note 10).

23. Quoted in "Panel Told All Landfills Leak," 1277.

24. "EPA Relaxes Hazardous Waste Rules," *Science*, 16 April 1982, 275–76.

25. "Panel Told All Landfills Leak," 1277.

26. Carol Scott (see note 10).

27. Quoted in Larry Waller, "Intel Sets Up Plants in Phoenix," *Electronics*, 12 April 1979.

28. Quoted in Monte Basgall, "Panelists Criticize Plans for Center," *Raleigh News and Observer*, 3 April 1981.

29. *Global Electronics Information Newsletter* (Pacific Studies Center), February 1983.

30. See Carlisle Ford Runge, "Positive Incentives for Pollution Control in North Carolina: A Policy Analysis," in *Making Pollution Prevention Pay*, ed. Donald Huisingh and Vicki Bailey (New York: Pergamon Press, 1982), 115–43.

31. G. A. Bulanowski et al., *A Survey and Analysis of State Policy Options to Encourage Alternatives to Land Disposal of Hazardous Waste* (Denver: National Conference of State Legislatures, July 1981).

32. See Donald Huisingh and Vicki Bailey, *Making Pollution Prevention Pay* (New York: Pergamon Press, 1982).

33. Michael Overcash, "Implication and Procedures for Waste Elimination of Hazardous Wastes," in Huisingh and Bailey, *Making Pollution Prevention Pay*, 72.

34. Michael G. Royston, "Making Pollution Prevention Pay," *Harvard Business Review* 6 (November-December 1980).

12 The Political Economy of Microelectronics in North Carolina

Paul Luebke, Stephen Peters, and John Wilson

The explosive growth of the microelectronics industry has attracted the attention of many economists interested in its organization and performance. One of the most intriguing aspects of this industry's growth is the nature of its geographical spread as different sectors are spun off from the original nucleus in California and Massachusetts to more peripheral regions within the United States and, eventually, to other countries. Economists have also paid attention to the efforts made by host regions and host nations to attract an industry so dynamic in its growth and seemingly environmentally unthreatening. What is largely missing from these accounts is the role of political factors in the growth and diffusion of microelectronics—what we have termed the political economy of microelectronics. To a large degree the changing economic organization of the industry is explained in terms of technological imperatives and economic efficiencies. Absent is any sense of conflict within host regions over the nature, speed, and direction of the growth of microelectronics within the region.

North Carolina has been the most industrial of the southeastern states for most of the twentieth century, and the state is widely known as the region's most "progressive." Yet the term "progressive" suffers from imprecise use, and in recent years it has been identified with politically liberal public policy. Using this definition, numerous academics and journalists examined North Carolina politics in the 1970s and concluded that the label was not justified.[1] North Carolinians who describe the state as "progressive," however, usually have a different meaning in mind. V. O. Key's classic 1949 essay on North Carolina politics captured this more restrictive sense of the term in its portrayal of North Carolina as a "progressive plutocracy": "forward-looking, yes, but always sound, always the kind of government liked by the big investor, the big employer." Although the forward-looking businessmen of the early and mid-twentieth century wanted to control government, they created a more active and centralized state than existed elsewhere in the South.[2]

From the 1920s into the 1970s, a "progressive" consensus appeared to unify the state's economic elites. State government provided major funding for both highway building and public education. In the 1920s, North Carolina became one of the first states to float a road-building bond issue "to get North Carolina out of the mud." The road-building issue was raised again in 1948, when an underdog gubernatorial candidate, Kerr Scott, won the election by promising rural voters that he would improve roads and pressure the power companies to extend electric lines to rural areas.[3] When local school boards were near bankruptcy during the Depression, the General Assembly passed a state sales tax and assumed responsibility for funding of the public schools. In the early 1960s, Governor Terry Sanford expanded state support for education by applying the sales tax to food and nonprescription drugs. Sanford's program illustrates Key's definition of "progressive," because an improved educational structure is part of the infrastructure necessary to diversify an industrial economy, while Sanford's choice of a regressive tax to fund the program illustrates how North Carolina state policy is often not "progressive" in a political sense.[4]

Luther Hodges, Sr., a former textile executive who was governor in the 1950s, actively supported private sector efforts to establish the Research Triangle Park, because he believed that such a center for research and development was necessary to expand the state's economy beyond its traditional textile–furniture low-wage base.[5] As late as the early 1970s, a "progressive" consensus prevailed on the need for an active state government to promote economic development through highway construction, education funding, and industrial recruitment. However, during the 1970s multinational corporations increased their interest in North Carolina as a haven from the higher-wage, highly unionized Midwest and Northeast, and the state's textile industry began to feel strong competition from imports. As a result the state's economic elites became divided on the wisdom of state government's active solicitation of higher-wage industry.[6] During the late 1970s, there were numerous documented cases in which local leaders of low-wage industry, mostly textiles, fought to prevent higher-wage, often unionized, firms from locating in their communities even when political officials and local workers wanted the new industry.[7]

In this essay we examine recent efforts to promote the microelectronics industry in North Carolina, in the light of both economic and political factors. We situate these efforts within the changing economic and political climate of North Carolina since the Second World War. Changing economic conditions certainly established the *preconditions* for the microelectronics industry, but two political factors have often been underestimated: first, the pattern of economic development in North Carolina has been shaped in part by specific and conscious political decisions; and second, economic and political elites disagree as to the wisdom of the initiatives to attract microelectronics. Governor James B. Hunt's commitment to bringing this industry to North Carolina was supported by moderate corporate leaders who advocate a more mixed industrial economic structure in North Carolina, if necessary at the expense of higher wages. Hunt's moderate position set him apart from conservatives within the state's Democratic party as well as conservative Republicans in the North Carolina General Assembly and U.S. Senate, who have remained lukewarm in their support for this kind of state entrepreneurship. We examine the economic bases from which these two groups operate, the nature of their interest, and the sources of their political support in order to account for the present state of the public effort to attract the microelectronics industry to North Carolina. We conclude with some speculation about the prospects of success for moderate corporate political leaders like Hunt in their quest for a wider mix of industries in the state.

Traditionalists and Modernizers

The willingness of parts of the microelectronics industry to leave California has given North Carolina government officials the hope that part of the industry can be attracted to the state. The General Assembly's decision to spend more than $40 million on the Microelectronics Center of North Carolina (MCNC) is both testimony to legislators' faith in the eventual success of their recruitment efforts and an additional reason for out-of-state microelectronics investors to select North Carolina. This government initiative, however, was neither the natural outgrowth of the state's present economic development nor was it undertaken with the consent of all the state's political leaders. In this section, we outline the political forces working for and against this initiative. Hunt and his supporters can be cast as "modernizers"; conservatives, including Republican U.S. Senator Jesse Helms, as "traditionalists." The modernizers are part of a large group that might be characterized as corporate liberals or, more accurately, as corporate pragmatists.

The position of corporate pragmatism represents the abandonment and rejection of the commitment to the incorporation of organized labor in economic decision making that was the chief hallmark of corporate liberalism.[8] It is different from traditional conservatism, however, in its strong support for the state's role as a promoter of private capital accumulation. To some degree corporate pragmatism can be seen as exploiting the weakness of the current union movement. Its willingness to concede relatively high wages along with unionization represents a marked departure from earlier corporate liberal strategies. For corporate pragmatists, North Carolina's special appeal is that the "high wages" offered by multinational corporations to nonunion workers are still about $2 an hour less than these firms would pay to unionized workers in the Northeast and Midwest or to nonunionized workers in California.

Governor Hunt viewed himself as a modernizer in the tradition established two decades ago by an earlier governor, Luther Hodges, Sr. He was a corporate pragmatist in the sense that while claiming to support the needs and aspirations of the state's poorer workers, he maintained a distance from labor unions.[9] This fits the pattern established by Hodges, who did not hesitate to call out the National Guard to break a textile workers' strike in Henderson in 1958, even though he supported economic development beyond the traditional textile and furniture sectors.[10] Hunt's strategy for improving the lot of the state's work force was to diversify the state's industrial base. Unlike more conservative politicians and business leaders in the state, Hunt argued that North Carolina should accommodate to changes in the national and international economy that render the traditional industries in the state (textiles, apparel, and furniture) vulnerable to competition from low-wage peripheral nations.

The corporate pragmatists have established political ties to the liberal wing of the national Democratic party (typified by Kennedy, Humphrey, and Mondale) in their efforts to promote economic diversification. These "pragmatists" include not only Hodges and Hunt but also nationally known figures such as former governor Terry Sanford and former U.S. Congressman Richardson Preyer. Within the state, this group has been mobilized by Bert Bennett, a wealthy Winston-Salem businessman who defines himself as a "moderate liberal." The term "moderate" applies to the group's economic development policies; its liberalism has centered on increased opportunities in state government for blacks and women. Under Hunt's governorship, blacks and women received political appointments and judgeships in unprecedented numbers. Although the corporate pragmatists enjoy wide support from many nonelite conservatives, their higher echelons are comprised chiefly of well-to-do cosmopolitan businessmen and professionals. Combining bases in the political offices of Raleigh, the university offices of Chapel Hill, Durham, and Raleigh, and business offices of Charlotte, Winston-Salem, Greensboro, and Raleigh, they share a conviction of the need for the state to steer and stimulate economic growth.

The more conservative forces in the state have increasingly aligned their sympathies with Jesse Helms and find their economic support in the traditional industrial strengths of North Carolina: textiles, apparel, and furniture. This group advocates a minimal role for government in economic affairs. It seeks to preserve the economic strength of the traditional industries, chiefly by protectionist measures and by vigorous opposition to unionism and government regulations that would raise the cost of doing business for smaller enterprises. It has been at best lukewarm in its support of industrial development efforts by the state, and often quite hostile.

During Hunt's administration conservative Democrats found themselves upstaged by the modernizers' rhetoric of better jobs and higher pay. To oppose Hunt's goals was politically dangerous, even if conservatives ideologically were opposed to such an activist development. The major concession that Hunt's industrial recruiters made to conservative Democrats who opposed higher-paying industry coming into their communities was to allow local elites to "redline" their areas.[11] In virtually all cases, the redlined areas were low-wage centers of textile, apparel, or furniture production.

Under these conditions opposition to the arrival of higher-wage industry in the state could not be direct; conservative Republicans led the indirect opposition by attacking the Hunt administration's zeal for "big government and high taxes." For example, Republican Beverly Lake correctly predicted in the 1980 gubernatorial election campaign that incumbent Hunt would raise gasoline taxes if reelected. Nevertheless, Hunt swamped Lake by a 62 percent to 38 percent margin while Reagan was upsetting Carter and an ultraconservative Republican (John East) was narrowly defeating a moderate-conservative Democrat incumbent U.S. senator. All indications in the early 1980s were that the Helms-dominated Republican party would continue its antigovernment attack against both Hunt and Democrat gubernatorial candidates to succeed him.

Recent Economic Developments in North Carolina

To understand why these political differences existed and why the strength of the corporate pragmatists increased so much at the expense of the traditional conservatives, it is necessary to look more closely at recent economic developments within the state. In the period since the Second World War, the economy of North Carolina has undergone dramatic change. Agriculture, although still vitally important to the state's economic well-being, has declined in importance as a money-earner and employer, to be replaced by manufacturing industries. Manufacturing in the state (which began in earnest at the turn of the century) was initially concentrated in the tobacco, textile, and furniture industries. By 1954 North Carolina ranked first in the nation in value of textiles, tobacco, and furniture produced. It ranked twelfth in total value added by manufacturing.

The initial postwar boom in manufacturing altered the economic structure of the state, but it failed to integrate North Carolina with the nation's economic mainstream. This early phase of industrialization was concentrated on semifinished goods, intended for "export" to other parts of the manufacturing economy. It was the least profitable, least skilled, and lowest paying of the stages of the production process.[12] However, a second wave of industrialization, beginning in the late 1960s, brought further changes in the state's economic structure.

North Carolina experienced tremendous industrial growth during the 1970s. Some of the traditional industrial leaders in the state were in the forefront of this expansion. Apparel manufacturing added more employers than any other industry between 1956 and 1977. Apparel plants, whose employees in 1970 were more than 80 percent female and who paid the lowest wage of any industry in the state, seemed to seek primarily female labor in rural areas but also tapped the "by-product" female labor generated by strong employment of males in cities. Tobacco steadily increased its level of automation, allowing the industry to rid itself of an earlier low-wage, labor-intensive character. By the late 1970s, tobacco had become very capital-intensive, with an increasingly high value added per employee.

While traditional industries like apparel, textiles, tobacco, and furniture continued to grow, they did not hold their proportionate share of the industrial base. A variety of industries, relatively new to North Carolina, began to take their place. For example, during the period between 1972 and 1977, there was a 217 percent increase in the value added by manufacture of fabricated metals. These firms were mostly manufacturing structural metal products for use in construction and were located in the urban Piedmont. Paper products more than doubled, as did the production of rubber and plastic goods. Both are industries in which plants tend to be large, capital-intensive, and high-wage-paying. The manufacture of electrical machinery and electrical equipment almost doubled. Chemicals increased by 82 percent. Although employment in this industry remained modest in comparison with many other states, the industry ranked high in value added per worker. Much of the industry concerned the manufacture of noncellulosic organic fibers, the rest producing industrial and agricultural chemicals, paint, and some pharmaceuticals.

Thus, by the late 1970s, although textiles still topped the list of annual value added by manufacture and tobacco ranked second, chemicals and electrical machinery were third and fourth.[13] As a result of these changes, a dualism, characteristic of the nation as a whole (in which North Carolina found itself relegated to the periphery), has been reproduced within the state itself, to the extent that core and peripheral sectors within the state can now be easily identified (see table 1).

Changes in the structure of the economy of North Carolina since the 1960s can be conceptualized in terms of a movement from the periphery of the nation's economy toward its core. Firms in core industries are noted for high productivity, high profits, intensive utilization of capital, high incidence of monopoly, and a high degree of unionization. Industries in the periphery are noted for their small firm size, labor intensity, low profit margins, low productivity, intensive product–market competition, lack of unionization, and low wages. Core industries include mining, construction, most forms of manufacturing, transportation, and finance. Peripheral industries include some kinds of manufacturing, wholesale and retail firms, services, and agriculture.[14] A look at thirty-year trends in the number of firms and number of employees within industries in North Carolina shows a shift in the balance of the state's economy away from a heavy reliance on peripheral industries toward a more even balance between core and periphery.

Employment in agriculture, a peripheral industry, has declined dramatically since the Second World War. In 1945, 40 percent of the labor force was employed on farms. By 1974, this proportion had fallen to 8 percent. As far as manufacturing is concerned, two measures of industrial growth can be used to compare 1947 with 1977. Measured in terms of the number of establishments in each industrial group, the greatest increases in the proportion of all manufacturing establishments occurred in machinery (up 6.5 percentage points), apparel (up 4.5), fabricated metals (up 3.7), and rubber products (up 2.5). Of these, machinery, fabricated metals, and rubber products are core industries.[15] Declines in proportion of all manufacturing establishments occurred in lumber and wood products (down 13.6 percentage points), textiles (down 36.0), and food and kindred products (down 6.4); only the last of these categories is a core industry. Measured in terms of number of employees, the greatest increases in employment have occurred in electrical machinery (up 3.9 percentage points), machinery (up 3.3), furniture (up 3.1), and rubber (up 2.8); all but furniture are core industries. Greatest declines have occurred in textiles (down 23.0 percentage points), tobacco manufacturing (down 5.5), and lumber and wood products; only tobacco manufacturing is part of the core. The North Carolina Commerce Department reported that "more than one of every three jobs announced in 1981 was involved in the manufacture of electrical machinery, fabricated metal products or transportation."[16]

Thus as far as manufacturing is concerned, typical core industries are beginning to be much more important to the economic well-being of the state. In 1947 peripheral industries accounted for 71 percent of the value added by manufacture in the state; by 1977 this proportion had fallen to 43 percent. The fastest-growing industries in North Carolina (measured by value added) between 1972 and 1977 were fabricated metals, paper production, rubber products, and electrical machinery.[17] In the year 1978–79, while the growth rate for all industries in the state was 2.4 percent, transportation equipment rose by 16.4 percent, machinery by 9.2 percent, meat packing by 7.7 percent, and primary metals by 6.8 percent.[18]

In the nonmanufacturing sector, the proportion of the nonagricultural labor force employed in finance, insurance, and real estate also rose, from 2.5 percent to 4.0 percent. Employment in wholesale and retail, in transportation, and in communications remained more or less constant. Great gains in employment were also being made in public administration. As a proportion of total nonagricultural employment, government employees increased from 12.0 percent in 1950 to 16.2 percent in 1977.[19]

Firms in core industries typically require a greater division of labor and tend to use more inputs than those in the periphery. The core sector is typified by capital-intensive firms in which value added per worker is high. Value added per worker in North Carolina is highest in the primary metal, chemical and allied, tobacco manufacturing, fabricated metal, and electrical machinery industries. The periphery is typified by firms in which the value added per worker is relatively low. In North Carolina, value added per worker is lowest in apparel, furniture, and textiles, the traditional mainstays of the North Carolina economy. The shift to more capital-intensive industries is indicated by the size of new investments: in 1969, only one new investment in the state exceeded $25 million; by 1977, the state could boast three $100 million investments in one year, and by 1978 two investments each totaling over $250 million.[20]

The long-term economic interests of capitalists in the core and pe-

Table 1. North Carolina Industry Characteristics, 1977

Industry	% All Value Added by Manufacture in State[a]	Size[b]
Food, kindred industry	5.4	7.3
Tobacco manufacturing	11.7	4.9
Textiles	32.0	27.0
Apparel	4.7	23.1
Lumber, wood	3.2	2.9
Furniture	6.7	20.8
Paper, allied industry	3.9	8.2
Printing, publishing	2.0	2.0
Chemical	8.9	7.7
Contract construction		
Transportation, utilities		
Stone, clay	2.4	3.9
Primary metal	1.5	9.8
Fabricated metal	5.4	6.9
Machinery	5.3	9.5
Wholesale, retail		
Finance, Information		
Service		
Government		

[a] Percentage of total value added in manufacturing contributed by industry group. Source: U.S. Dept. of Commerce, Census of Manufactures, 1977 (Washington, D.C., 1981).

[b] Percentage of establishments in that industry group with 240 or more employees. Source: North Carolina Abstracts, 1979, 53.

[c] Percentage of all nonagricultural employees by industry group. Source: North Carolina Abstracts, 1979, 321.

[d] Hourly wage for production workers, exclusind overtime, averaged over the year. Source: North Carolina Abstracts, 1979, 32.

ripheral industries are not alike in all respects, especially with regard to the role of the state in economic affairs. Firms in the periphery face considerable competition, which forces wages down. Owners of firms in the periphery thus typically oppose unions, welfare plans, and government programs that would increase their taxes and reduce their small profit margins. Being less capital-intensive, peripheral firms have less organization to protect and less need for the trained personnel the state might provide.[21] Firms in the core, on the other hand, are better

% of Employees[c]	Wage[d]	Value Added per Worker[e]	Concentration Ratio[f]	Profit[g]
1.9	3.87	25,069	21	4.2
1.2	5.78	91,904	71	10.4
12.2	3.89	17.417	71	4.7
4.0	3.13	11.063	0	4.8
2.0	3.58	18,243	0	6.4
3.5	3.84	16,660	0	5.4
0.9	5.90	37,668	6	6.4
0.8	4.62	24,146	27	8.9
1.7	5.15	50,393	26	8.9
4.8	NA			3.5
4.8	NA			6.1
0.8	4.54	30,873	30	6.7
0.3	4.85	55,560	27	2.6
8.0	4.87	35,684	8	6.7
0.7	4.62	29,039	7	10.5
19.4	4.07			2.8
3.9	NA			7.9
13.8	NA			5.0
16.2	NA			

[e]Total value added by manufacture in that industry group. Source: North Carolina Abstracts, 1979, 32.

[f]Percentage of industries in that industry group whose concentration ratio is 60% or above, where the concentration ratio is a measure of the amount of total shipment value accounted for by the four largest firms in the industry. Source: U.S. Dept. of Commerce, Census of Manufactures, 1977 (Washington, D.C., 1981).

[g]Net income expressed as a percentage of gross income. Source: IRS Corporate Income Tax Returns, 1977: Statistics of Income, 18-20.

placed to determine their own prices and to pass on wage increases and tax increases to their buyers. They tend to be preoccupied with longer-term expansion rather than short-term market share. Unlike the peripheral firms, which tend to regard the intervention of the state into the economy as a damaging assertion of power, the core firm sees state management of the economy as an accommodation by the state to more exact and advanced technologies. "The fully planned economy, so far from being unpopular, is warmly regarded by those who know it best."[22] Core firms thus favor state policies that preserve the conditions for a steady accumulation of capital and the integrity of the fiscal and monetary institutions within which their security is rooted. The state is also expected not only to protect the infrastructure but to provide social welfare programs to ameliorate labor–capital conflict.

The Economic Bases of Political Action

The presumption behind campaign contributions is that donors gain access to successful candidates. We therefore supposed that employees of core and peripheral firms would prefer different candidates and, further, that candidates would receive differing proportions of their contributions from the various industrial sectors. Specifically, we hypothesized that donors from the core would have contributed disproportionately to campaigns of modernizers such as Governor Hunt or former Charlotte banker Luther Hodges, Jr., who ran unsuccessfully for the Democratic U.S. Senate nomination in 1978, whereas contributors from peripheral industries such as textiles and furniture would have disproportionately supported a traditionalist like Senator Helms. We tested this hypothesis by analyzing records of contributions to Hunt's campaigns for lieutenant governor in 1972 and governor in 1976, Hodges's sole run for the U.S. Senate in 1978, and Helms's first two races for the Senate, in 1972 and 1978.[23]

Although both Helms and Hodges attracted considerable financial support from outside the state, our analysis was limited to in-state contributions because of our interest in the link between in-state business support and political action. The records of campaign donations include the name of the contributor and the amount of the contribution, and in most cases the contributor's occupation and place of employment. Because the economic basis of the contribution was our chief interest, we further limited our analysis only to data for those contributors whose occupation and firm (where appropriate) could be identified. Often the nature of the firm (for example, textile mill, or bank) was clear from the list itself; otherwise this information was traced through the *North Carolina Directory of Manufacturing Firms* or through telephone directories.[24] Within these restrictions ultimately 50.6 percent of the contributors and 74.4 percent of the money given to Hunt's campaigns could be traced; 41.4 percent of the contributors and 55.4 percent of the money given to Helms; and 63.3 percent of the contributors and 83.3 percent of the money given to Hodges. Unfortunately there is no means of checking for biases resulting from the eliminated (incomplete) data.

The purpose of identifying the kind of firm for which each contributor worked was to identify the industry group to which it belonged and, in turn, its location in the core or the periphery of the national economy. For our analysis we classified these firms according to a table devised for a study of industrial segmentation,[25] which categorizes industries into fourteen major groups (for example, construction, manufacturing, transportation, wholesale trade), each further subdivided into several more specific industries (manufacturing, for example, includes twenty-nine industries such as lumber and wood products, electrical machinery, and printing). This detailed classification allowed us to locate each firm within a specific industry and each specific industry within an industry group. A thorough analysis would require more information on firms than was available, at least for the purposes of classification into core or periphery. For example, the table we used classifies some kinds of wholesaling as core (such as drugs, chemicals, electrical goods, machinery, and alcoholic beverages) and some as peripheral. The data on hand did not always permit such a clear identification. Although some of the wholesaling firms on our list may fall into those "core" industries, we classified all wholesaling and retailing as peripheral. The classification table places all manufacturing in the core with the exception of textiles, lumber and wood products, furniture, miscellaneous plastic products, and leather products. We accordingly treated the first three of these five industries as a separate peripheral category (the latter two did not occur in our lists). The distinctive treatment of wholesale and retail trade, and these selected kinds of manufacturing, as peripheral may have skewed the distribution of firms in our statistics toward an overcounting of peripheral firms.

The summary tables presented here highlight certain industrial categories within both core and periphery. Among core industries we singled out construction and transportation because those industries seemed to occur with great frequency on Hunt's contribution list; public administration, because state officials contributed heavily to Hunt's campaign; finance (banking, credit agencies, security brokerage, insurance; real estate was classified as peripheral), because of Hodges's background; and professional and related services. "Other core" includes manufacturing (other than textiles, lumber, and furniture), communications, and utilities. Among peripheral industries we singled out textiles (combined with apparel), furniture (combined with lumber and wood products), and wholesaling and retailing. "Other periphery" consists largely of agriculture, business and repair services, and personal services. We treated separately, as a third group, contributors who listed their occupation as housewife.

The tables show that our hypothesis was supported. As seen in table 2, employees of core corporations were more likely to contribute to the modernizing candidates, Hunt and Hodges (64 percent and 69 percent) than they were to the traditionalist, Helms (47 percent). Helms received disproportionate support from employees of peripheral firms (46 percent, compared to Hunt's 36 percent and Hodges's 28 percent). An examination of contributors from two key peripheral industries, textiles and furniture, reveals a similar pattern: 22 percent of Helms's were from those industries, in contrast to 5 percent of Hunt's and 6 percent of Hodges's. Table 3 analyzes dollar amounts, rather than contributors, from core and peripheral firms. More than 50 percent of Hunt's and Hodges's contributions came from the core, but less than 40 percent of Helms's. The textile and furniture industries show a striking preference for Helms: 40 percent of his funding came from these two peripheral industries, compared to 5 percent of Hunt's and 4 percent of Hodges's.

No doubt because of his ties to North Carolina's banking industry, Hodges received a disproportionate amount of his funding (26 percent) from the financial sector. Perhaps because access to the governor's office is particularly important for construction and transportation firms, 20 percent of Hunt's contributions were raised from those industries, in sharp contrast to the funding pattern for the U.S. Senate candidates, Helms and Hodges (3 percent and 6 percent).

Overall, these tables provide the basis for a more general assessment of sources of support for moderate and conservative politicians. Hunt's and Hodges's strength lay in core industries like construction and transportation (Hunt) and finance (Hodges); both also received much support from the real estate industry, classified in "Other Peripheral" here but having an obvious interest in the modernizing aspects of property development. Helms was strong in textiles and furniture (both furniture per se and lumber as a subdivision). In sum, those with the most to gain from "growth" (and the least to fear from wage competition) were more likely to support politicians who favored state activism as a catalyst to private-sector economic development. The pattern of campaign contributors, together with speeches and declarations from members of the two groups concerned, confirms what we expected of the political support and political behavior of modernizers

Table 2. Distribution of Campaign Contributors to Each Candidate across Industrial Sectors (percentage)

Sector	Hunt 1972, 1976	Helms 1972, 1978	Hodges 1978
Core			
Construction	9.9	3.8	5.1
Transportation	6.0	6.8	2.7
Finance	11.7	7.2	22.8
Public administration	8.3	0.0	1.4
Professionals	18.8	25.2	23.9
Other core	9.3	10.2	12.6
Core subtotal	(64.0)	(47.2)	(68.5)
Periphery			
Textiles	4.2	15.0	3.8
Furniture	1.2	6.9	2.1
Wholesale and retail	15.1	11.9	10.8
Other periphery	15.4	12.5	10.8
Periphery subtotal	(35.9)	(46.3)	(27.5)
Housewives	0.1	6.4	4.0
TOTAL CONTRIBUTORS	652	234	569

and traditionalists. The influence of the modernizers increases as the state's dependence on peripheral industries diminishes.

Despite a national swing to the right, the state's core industries continued to support Hunt's moderate policies. Hunt sought both to strengthen this base in the core sector and to broaden his appeal, in part by his politically astute packaging of the microelectronics initiative. After his victory in the 1976 election he actively solicited support from conservative Democratic businessmen who had opposed him in the Democratic primary, by organizing a series of economic development conferences. At the first of these, in January 1978, he proposed lowering the manufacturer's inventory tax. Subsequent conferences were designed to confirm his image of being pro-business. Hunt successfully avoided serious Democratic challenges in both his 1980 bid for reelection as governor and his 1984 race for the U.S. Senate.

The only organized opposition to Hunt's corporate pragmatism has come from the state's Republicans, dominated by Jesse Helms. Helms's

Table 3. The Distribution of Campaign Contributions for Each Candidate across Industrial Sectors (percentage)

Sector	Hunt 1972, 1976	Helms 1972, 1978	Hodges 1978
Core			
Construction	14.4	2.9	4.2
Transportation	6.0	0.3	2.1
Finance	7.4	5.8	26.0
Public administration	5.9	0.0	0.4
Professionals	10.3	17.7	17.7
Other	11.7	11.7	2.3
Core subtotal	(55.7)	(38.4)	(52.7)
Periphery			
Textiles	3.2	31.6	2.3
Furniture	1.7	8.1	1.8
Wholesale and retail	15.1	11.9	14.3
Other periphery	24.2	5.3	25.7
Periphery subtotal	(44.2)	(56.9)	(44.1)
Housewives	0.1	4.7	3.2
TOTAL DOLLARS	$771,971	$154,040	$328,490

support from the peripheral sector seems secure. On behalf of industries seriously threatened by imports (often from foreign subsidiaries of U.S. multinationals) Helms has supported protectionist legislation as well as opposed "big government." Representatives of the more traditional industries have felt excluded from Hunt's modernizing campaigns. One spokesman complained that "our policy makers have at times lost sight of the fact that along with the new industrial growth, the state's textile industry, the state's largest industry and employer of manufacturing workers, has been here all along and provided a tremendous industrial base upon which the state's economy could grow." He urged that legislators "develop and maintain the attitude that existing industries are just as important as new industries."[26]

The traditionalists, however, do not directly address the issue of higher wages for the state. Rather they seek to shift the debate from the virtue of specific corporate pragmatist proposals (for instance, the promotion of industrial efficiency and thus real income growth) to the more abstract benefits of a small state sector that does not impinge upon the businessman's traditional entrepreneurial freedom. The debate over the Microelectronics Center fits exactly in the middle of this conflict. Hunt and his supporters saw it as a vital part of their modernizing campaign. Helms and other Republicans saw it as a bad case of state planning, the impingement of public regulations on private freedoms, and favoritism toward out-of-state corporations at the expense of North Carolina's traditional industries.

Conclusion

During his two terms as governor, Jim Hunt immeasurably strengthened the corporate pragmatist perspective within the North Carolina Democratic party. He placed conservative business-oriented Democrats on the defensive and, through political appointments of blacks and women, protected his left flank from urban liberals and populists who saw him as too friendly with the politically moderate corporate establishment. If our analysis is correct, the corporate pragmatist perspective will be adopted by any Democrat who seeks the governorship in upcoming elections in the 1980s and 1990s. As a correlate, if enough higher-wage, high-technology industry related to microelectronics is attracted to North Carolina, a more conservative Democratic politician in the 1980s would be unlikely to risk attacking the Microelectronics Center of North Carolina. Plausibly, the state-government–corporation–university relationship that is fundamental to corporate pragmatism will go unchallenged by conservative Democrats, even though they object to "activist government."

Republicans are less likely to support initiatives like the Microelectronics Center, the more conservative of them arguing that the corporations seeking to move to lower-wage, less unionized regions will carefully consider coming to North Carolina anyway, as long as a "favorable business climate" is maintained. The 1984 gubernatorial election illustrated this difference between the corporate pragmatists in the Democratic party and the traditionalists among the Republicans. Jim Martin, a Republican congressman from the Charlotte area, won that election by a margin of 54 percent to 46 percent.

Martin is the state's second Republican governor in the twentieth century. His twelve-year congressional voting record was similar to that of Senator Helms, whose victory over Hunt was less decisive (51 percent to 48 percent). However, Martin, a former chemistry professor from Davidson College, successfully projected an urbane, modern image that contrasted with the images of both Senator Helms and Martin's Democratic opponent. Martin's campaign soft-pedaled his conservative past and hinted at a commitment to reasoned pragmatism. As a result, in marked contrast to his fellow Republican Helms, Martin received relatively strong support from urban, middle-class precincts across the state whose voters are oriented to the high-technology and core-industry sectors of North Carolina's economy.

The first months of the Martin administration suggested that Martin's economic development policy would attempt to balance the interests of both core and peripheral sectors. Symbolically, Martin hired a peripheral-sector representative, the furniture industrialist Howard H. Haworth, as his secretary of commerce. Underscoring his belief—and that of peripheral-sector Republican businessmen—that the Democrats under Hunt had neglected North Carolina's homegrown industries in favor of out-of-state corporations, Martin created the position of Assistant Commerce Secretary for Traditional Industries.

At the same time, the microelectronics initiatives of the Hunt administration appear to have become institutionalized. Martin stressed in public comments that he would support both new (microelectronics) and traditional business. His proposal to remove the state tax on manufacturers' inventories, while immensely popular among peripheral-sector industrialists, would also, if successful, benefit core-sector businessmen with large on-hand inventories. Martin appeared to be steering a pragmatic, if generally conservative, course that would be less partial toward out-of-state core-sector industries than the policies of his Democratic predecessor.

Notes

1. Paul Luebke, "Corporate Conservatism and Government Moderation in North Carolina," in *Perspectives on the American South*, ed. Merle Black and John Shelton Reed (New York: Gordon & Breach, 1981), 107–8, 116–17.

2. V. O. Key, Jr., *Southern Politics in State and Nation* (New York: Knopf, 1949), 213–14.

3. Thomas Parramore, *Express Lanes and Country Roads: The Way We Lived in North Carolina, 1920–1970* (Chapel Hill: University of North Carolina Press, 1983), 73–74.

4. See Robert Goldman and Paul Luebke, "Toward a Corporate Pragmatism: The Rise and Demise of the North Carolina Labor Center" (1983, typescript); Robert Goldman and Paul Luebke, "Corporate Capital Moves South: Competing Class Interests and Labor Relations in North Carolina's 'New' Political Economy," *Journal of Political and Military Sociology* 13 (Spring 1985); and Luebke, "Corporate Conservatism and Government Moderation."

5. Luther Hodges, *Businessman in the State-House: Six Years as Governor of North Carolina* (Chapel Hill: University of North Carolina Press, 1962), chap. 9; Paul Luebke and Joseph Schneider, "Economic and Racial Ideology in the North Carolina Elite" (Paper delivered at the 1983 meetings of the Southern Sociological Society, Atlanta).

6. Luebke, "Corporate Conservatism and Government Moderation," 110–11.

7. Paul Luebke, Bob McMahon, and Jeff Risberg, "Selective Industrial Recruitment in North Carolina," *Working Papers for a New Society* 6 (March–April 1979): 14–17.

8. Goldman and Luebke, "Toward a Corporate Pragmatism"; James Weinstein, *The Corporate Ideal in the Liberal State* (Boston: Beacon Press, 1968).

9. Paul Bernish, "Unionism Is Hot Political Issue in N.C. Campaign," *Charlotte Observer*, 19 October 1976.

10. Hodges, *Businessman in the State-House*, chap. 10.

11. Luebke, McMahon and Risberg, "Selective Industrial Recruitment."

12. Hugh Lefler, *History of North Carolina*, 2 vols. (New York: Lewis Historical Publishing Co., 1956), 2:115.

13. For these figures, and those given in the two paragraphs preceding, see Charles Heatherly, "North Carolina's Changing Industrial Scene," *North Carolina Commerce Report* 2 (Winter 1980): 12–13.

14. See Charles Tolbert, Patrick Horan, and E. M. Beck, "The Structure of Economic Segmentation: A Dual Economy Approach," *American Journal of Sociology* 85 (1980): 1095–1116.

15. "General Statistics and Distribution of Establishments by Employment Size Groups, by Industry for the State" (Table 4), in U.S. Department of Commerce, Bureau of the Census, *Census of Manufactures, 1947* (Washington, D.C., 1950), vol. 3; "Statistics by Selected Industry Groups and Industry for the State, 1977 and 1972" (Table 5), in idem, *Census of Manufactures, 1977* (Washington, D.C., 1981), vol. 3.

16. "Industrial Development Tops $2 Billion Again in 1981," *North Carolina Commerce Report* 4 (Spring 1982): 3.

17. Heatherly, "North Carolina's Changing Industrial Scene," 12.

18. "High Wage Gains," *North Carolina Commerce Report* 1 (Summer 1979): 7.

19. "Estimated Annual Average Employment in Nonagricultural Establishments and Average Hourly Earnings," *North Carolina State Government Statistical Abstract* (Raleigh: Division of State Budget and Management, 1979), 192–93.

20. Heatherly, "North Carolina's Changing Industrial Scene," 13.

21. John Kenneth Galbraith, *The New Industrial State* (Boston: Houghton Mifflin, 1971), 305.

22. Galbraith, *The New Industrial State*, 31.

23. Thad Beyle, of the Department of Political Science at the University of North Carolina at Chapel Hill, generously allowed us to use his data for these campaigns.

24. *North Carolina Directory of Manufacturing Firms* (Raleigh: North Carolina Department of Labor, 1979).

25. Tolbert, Horan, and Beck, "The Structure of Economic Segmentation," 1110.

26. Jerry Roberts, "Textiles Provide Sound Base for North Carolina Economy," *North Carolina Commerce Report* 1 (Winter 1979): 6.

Contributors

Paul S. Adler is an assistant professor of engineering management at Stanford University. He received his doctorate in economics and management from the University of Picardy (France). He was a research economist for the French Ministry of Labor, and before coming to Stanford was a postdoctoral research fellow at the Harvard Business School. His principal research interests are in technology management and productivity performance in engineering, manufacturing, and service operations.

Tom Bourgeois is an economist with the Ways and Means Committee, State Capitol, Albany, New York, the State Assembly's primary fiscal staff. He received his undergraduate degree in economics at Union College and a master's in regional planning at the University of North Carolina at Chapel Hill. He spent two semesters in the Ph.D. program in policy analysis at the Rand Corporation in Santa Monica, California. Subsequently he has worked in the area of economic forecasting, tax revenue estimating, and tax policy analysis with the New York State Division of Budget and in his current position.

Harvey Goldstein is an associate professor in the Department of City and Regional Planning at the University of North Carolina at Chapel Hill. He previously taught at Columbia University and Virginia Polytechnic Institute. His Ph.D. was earned at the University of Pennsylvania in city and regional planning. He has consulted and conducted policy research during the past five years for the U.S. Bureau of Labor Statistics on projecting industrial employment change in state and substate areas. Current areas of research interest include state industrial policy design, regional business cycles, and regional economic restructuring.

Rosalind Greenstein is a doctoral candidate in the Department of City and Regional Planning at the University of North Carolina at Chapel Hill. The working title of her dissertation is "Metropolitan Economic Structure and Labor Market Outcomes: Economic Inequality in a Changing International Economy." Her research interests include community economic development, labor market outcomes of economic development, and the organization of work. She received a B.A. in economics from the University of California at Santa Cruz.

John S. Hekman is an associate professor in the School of Business at the University of North Carolina at Chapel Hill. He holds a B.A. in history from Valparaiso University and an M.B.A. in finance from the University of Chicago. He received his Ph.D. in economics from Chicago and has worked on the research staff at the Federal Reserve Bank of Boston and taught economics at Boston College. His research interest is focused mainly in real estate and urban and regional development.

Paul Luebke is an associate professor of sociology at the University of North Carolina at Greensboro. He has written extensively on economic development, social movements, and politics in contemporary North Carolina.

Michael I. Luger is an assistant professor of public policy studies and economics at Duke University. He received a bachelor's degree in architecture and urban planning and a master's in public and international affairs from Princeton University, and a master's degree in city and regional planning and a Ph.D. in economics from the University of California, Berkeley. In 1983–84 he served on a committee of Governor Hunt's Task Force on Science and Technology and as a member of a White House Task Force on Productivity. Since 1981 he has been on the board of directors of a Durham-based development corporation. His research and teaching interests include state and local economic development, federal tax policy, and macroeconomic policy.

Emil E. Malizia is an associate professor of city and regional planning at the University of North Carolina at Chapel Hill. He has taught the theories, strategies, and techniques of economic development to graduate students and practicing professionals since 1969. He also has extensive policy research experience and numerous publications on urban, rural, and regional development. He has been a senior Fulbright scholar in Colombia, South America, and a special assistant in the Employment and Training administration of the U.S. Department of Labor. Most recently, he has designed and initiated economic development projects as a consultant to businesses, developers, and governments in the Southeast. He holds an M.R.P. and Ph.D. from Cornell University.

Stephen Peters recently completed his Ph.D. in sociology at Duke University. His dissertation was on the development of local capitalist classes in Chile and Nigeria during the 1960s and 1970s. He is presently employed at Duke's Talent Identification Program and is conducting research at the university.

F. Dana Robinson received his Ph.D. in public policy analysis from the Wharton School of the University of Pennsylvania. He is currently an analyst with First Pennsylvania Bank in Philadelphia.

Carlisle Ford Runge is an assistant professor of agricultural and applied economics and an adjunct member of the Hubert H. Humphrey Institute of Public Affairs at the University of Minnesota. He is a former member of the political science faculty at the University of North Carolina at Chapel Hill. He was an undergraduate at Chapel Hill, received his M.A. in politics and economics as a Rhodes Scholar at New College, Oxford University, and gained his Ph.D. in agricultural economics at the University of Wisconsin. In 1982 he was elected a Science and Diplomacy Fellow of the American Association for the Advancement of Science. His research focuses on natural resources policy, agricultural policy, and land use.

Gregory B. Sampson is Research Director at the North Carolina Employment Security Commission. He did his undergraduate work at William and Mary in the social sciences and holds master's degrees in demography and regional planning from Brown University and the University of North Carolina, respectively. He received his Ph.D. in sociology from the University of North Carolina. From 1983 to 1985 he served as Executive Director of the Governor's Oversight Committee for Official Labor Market Information. Currently he is conducting applied research on the structure and operation of local labor markets and working on forecasting models of employment changes.

James I. Stein is a doctoral student in the Department of Urban Studies and Planning at the Massachusetts Institute of Technology. He received his B.A. in economics from Stanford University and his master's degree in regional planning from the University of North Carolina at Chapel Hill. His principal research interest is in the effects of corporate restructuring on development policies in the Third World.

Dale Whittington is an assistant professor with the Department of City and Regional Planning at the University of North Carolina at

Chapel Hill. He did his undergraduate work at Brown University in economics and holds master's degrees from the Lyndon B. Johnson School of Public Affairs and the London School of Economics and Political Science. He received his Ph.D. in business, engineering, and public affairs from the University of Texas at Austin. He has served as staff economist with the National Commission on Water Quality and as systems analyst with the Egyptian Academy of Scientific Research and Technology. In 1983 he was a Fulbright scholar at the Institute of Environmental Studies at the University of Khartoum (Sudan). He is the author of *Water Management Models in Practice: A Case Study of the Aswan High Dam* (1983).

John Wilson is an associate professor with the Department of Sociology at Duke University. He received his D.Phil. in sociology from the University of Oxford. His most recent book, *Social Theory*, appeared in 1983. His major current research interests are an assessment of the impact of off-farm work among 697 farm families in North Carolina (with Ida Simpson) and a study of land ownership patterns in North Carolina (with the Institute for Southern Studies).

Index